Communications System Engineering Handbook

Communications System Engineering Handbook

Contributors

A. Ghassemi and T. A. Gulliver et al.

AURIS
Reference

www.aurisreference.com

Communications System Engineering Handbook

Contributors: A. Ghassemi and T. A. Gulliver et al.

Published by Auris Reference Limited
www.aurisreference.com

United Kingdom

Communications System Engineering Handbook

ISBN: 978-1-78154-818-9

British Library Cataloguing in Publication Data
A CIP record for this book is available from the British Library

Printed in the United Kingdom

Exclusively distributed by CBS Publishers & Distributors Pvt. Ltd.

Sales & Distribution Rights only for India, Pakistan, Bangladesh, Sri Lanka, Nepal and Bhutan. This book is not to be sold outside these territories.

Contents

List of Abbreviations

ACR	Absolute Category Rating
ADR	Active degenerative resistance
AODV	Ad hoc On-Demand Distance Vector
AOA	Angle of Arrival
APD	Avalanche photodiode
CAP	Contention Access Period
CFP	Contention Free Period
CT	Continuous-time
DN	Destination Node
DBPSK	Differential Binary Phase Shift Keying
DSP	Digital signal processors
DD	Direct detection
DHT	Distributed Hash Table
EMG	Electromyography
EDFA	Erbium-doped fiber amplifier
EKF	Extended Kalman Filter
FIR	Finite impulse response
GUI	Graphical User Interface
GTS	Guaranteed Time Slots
ION	Initial Overlay Node
IM	Intensity modulation
IMM	Interacting Multiple Model
IN	Intermediate Node
ITU	International Telecommunication Union
JN	Joining Node
LS-MDS	Least Square-Multidimensional Scaling
LMS	Least-mean-squares
LED	Light Emitter Diode
LOS	Line-of-sight
MAE	Mean absolute error
MDS	Multidimensional Scaling
MPC	Multi-path components
OSM	Open Street Map
OPLL	Optical Phase Lock Loop
OWC	Optical Wireless Communications
PLE	Path loss exponent
PDP	Power delay profile
RFID	Radio frequency identification
RF	Radio link
RSSI	Received Signal Strength Indication
SOA	Semiconductor optical amplifiers

SIP	Session Initiation Protocol
SFG	Signal flow graph
SNR	Signal to noise ratio
SOP	State of polarization
SAW	Surface acoustic wave
TMJ	Temporomandibular joint
TOA	Time of Arrival
TTL	Time to Live
UWB	Ultra wideband
V2V	Vehicle-to-vehicle
VC	Videoconference
WDM	Wavelength division multiplexing
WON	Wireles optical networks
WSN	Wireless Sensor Networks

List of Contributors

Ghassemi
Department of Electrical and Computer Engineering, University of Victoria, Victoria, BC, Canada

T. A. Gulliver
Department of Electrical and Computer Engineering, University of Victoria, Victoria, BC, Canada

C. Tselios
Wireless Telecommunications Laboratory, University of Patras, TK 26504, Patras, Greece

C. Papageorgiou
Wireless Telecommunications Laboratory, University of Patras, TK 26504, Patras, Greece

K. Birkos
Wireless Telecommunications Laboratory, University of Patras, TK 26504, Patras, Greece

I. Politis
Wireless Telecommunications Laboratory, University of Patras, TK 26504, Patras, Greece

T. Dagiuklas
Department of Telecommunications Systems and Networks, TEI of Messolonghi, TK 30300, Nafpaktos, Greece

E. Barka
Information Security Track, College of Information Technology, United Arab Emirates (UAE) University, Al Ain, United Arab Emirates

A. Lakas
Networking Track, College of Information Technology, United Arab Emirates (UAE) University, Al Ain, United Arab Emirates

Aohan Li
School of Electrical Engineering, Heilongjiang University, Harbin 150800, China

Ziheng Yang
School of Electrical Engineering, Heilongjiang University, Harbin 150800, China

Renji Qi
School of Electrical Engineering, Heilongjiang University, Harbin 150800, China

Feng Zhou
School of Electrical Engineering, Heilongjiang University, Harbin 150800, China

Guangjie Han
Department of Information and Communication System, Hohai University, Changzhou 213022, China

Andres Diaz Lantada
Mechanical Engineering Department at Universidad Politécnica de Madrid, c/José Gutiérrez Abascal 2, 28006 Madrid, Spain

Carlos González Bris
Conectivity Group, Higher Technical School of Telecommunication Engineering, Universidad Politécnica de Madrid, Av. Complutense 30, 28040 Madrid, Spain

Pilar Lafont Morgado
Mechanical Engineering Department at Universidad Politécnica de Madrid, c/José Gutiérrez Abascal 2, 28006 Madrid, Spain and Jesús Sanz Maudes

Conectivity Group, Higher Technical School of Telecommunication Engineering, Universidad Politécnica de Madrid, Av. Complutense 30, 28040 Madrid, Spain

Juan Chóliz
Research Institute of Engineering in Aragón, I3A, University of Zaragoza, C/María de Luna 3, Zaragoza 50018, Spain

Ángela Hernández
Research Institute of Engineering in Aragón, I3A, University of Zaragoza, C/María de Luna 3, Zaragoza 50018, Spain

Antonio Valdovinos
Research Institute of Engineering in Aragón, I3A, University of Zaragoza, C/María de Luna 3, Zaragoza 50018, Spain

Gao Zhiqiang, Associate Professor
Department of microelectronics of Harbin Institute of Technology China

Juan-de-Dios Sánchez- López
Autonomus University of Baja California

Arturo Arvizu M
Cicese Research Center México

Francisco J. Mendieta
Cicese Research Center México

Iván Nieto Hipólito
Autonomus University of Baja California

Robert Nagel
Institute of Communication Networks, Technische Universität München Germany
Stefan Morscher
Institute of Communication Networks, Technische Universität München Germany

Felicitas Hesselmann
Institute for Research Information and Quality Assurance, Schützenstraße 6a, 10117
Berlin, Germany

Verena Wienefoet
Institute for Research Information and Quality Assurance, Schützenstraße 6a, 10117
Berlin, Germany

Martin Reinhart
Institute for Research Information and Quality Assurance, Schützenstraße 6a, 10117
Berlin, Germany
Humboldt-Universität zu Berlin, Institut für Sozialwissenschaften, Unter den Linden
6, 10099 Berlin, Germany

Mohammed H. Alsharif
Department of Electrical, Electronics and Systems Engineering, Faculty of Engineering and Built Environment, Universiti Kebangsaan Malaysia, 43600 Bangi, Selangor,
Malaysia

Rosdiadee Nordin
Department of Electrical, Electronics and Systems Engineering, Faculty of Engineering and Built Environment, Universiti Kebangsaan Malaysia, 43600 Bangi, Selangor,
Malaysia

Mahamod Ismail
Department of Electrical, Electronics and Systems Engineering, Faculty of Engineering and Built Environment, Universiti Kebangsaan Malaysia, 43600 Bangi, Selangor,
Malaysia

Preface

A communications system is a collection of individual communications networks, transmission systems, relay stations, tributary stations, and data terminal equipment usually capable of interconnection and interoperation to form an integrated whole. The text Communications System Engineering Handbook provides an exposition of the theory for communications system. First chapter provides an overview of the major distortionless PAPR reduction techniques, namely, partial transmit sequence (PTS), selective mapping (SLM), and tone reservation (TR). Second chapter describes the design and prototype implementation of a communication platform aiming to provide voice and video communication in a distributed networking environment. In third chapter, we introduce a new concept of access control for enhancing the security of the SIP-based communications. Fourth chapter focuses on the code synchronization problem in spread spectrum (SPSP) communications. In fifth chapter, we propose a system and explain the development of a prototype that uses a wireless connection between a passive force sensor, located within a splint in the mouth cavity, and an active external unit that energizes the sensor and permanently records all force measurements. Sixth chapter provides a complete system level evaluation of a UWB-based communication and location system for wireless sensor networks, including aspects such as UWB-based ranging, tracking algorithms, latency, target mobility and MAC layer design. The design of CMOS integrated q-enhanced RF filters for multi-band/mode wireless applications has been outlined in seventh chapter. Eighth chapter focuses on trends of the optical wireless communications. Ninth chapter deals with connectivity prediction in mobile vehicular environments backed by digital maps. Tenth chapter draws on research traditions and insights from criminology to elaborate on the problems associated with current practices of measuring scientific misconduct. Last chapter presents an overview of the energy consumption problems of wireless communication networks and describes the techniques that have been used to improve the energy efficiency of these networks.

Chapter 1

LOW-COMPLEXITY DISTORTIONLESS TECHNIQUES FOR PEAK POWER REDUCTION IN OFDM COMMUNICATION SYSTEMS

Ghassemi and T. A. Gulliver

Department of Electrical and Computer Engineering, University of Victoria, Victoria, BC, Canada

ABSTRACT

A high peak-to-average power ratio (PAPR) is one of the major drawbacks to using orthogonal frequency division multiplexing (OFDM) modulation. The three most effective distortionless techniques for PAPR reduction are partial transmit sequence (PTS), selective mapping (SLM), and tone reservation (TR). However, the high computational complexity due to the inverse discrete Fourier transform (IDFT) is a problem with these approaches. Implementation of these techniques typically employ direct computation of the IDFT, which is not the most efficient solution. In this paper, we consider the development and performance analysis of these distortionless techniques in conjunction with low-complexity IFFT algorithms to reduce the PAPR of the OFDM signal. Recently, proposed IFFT-based techniques are shown to substantially reduce the computational complexity and improve PAPR performance.

INTRODUCTION

Multicarrier modulation is a data transmission technique, which provides efficient bandwidth utilization and robustness against time dispersive channels. Examples of multicarrier modulation systems are digital audio broadcasting (DAB), digital video broadcasting (DVB), and wireless local and metropolitan area networks using orthogonal frequency division multiplexing (OFDM), and digital subscriber line (DSL) using discrete multitone (DMT) systems. OFDM is an effective transmission technique for wireless communications over frequency selective channels as it provides immunity to multipath fading. An inverse fast Fourier transform (IFFT) and a fast Fourier transform (FFT) are typically employed for baseband modulation and demodulation, respectively. Using an IFFT/FFT simplifies the design of the transceiver and eliminates the need for high speed equalizers, resulting in an efficient hardware implementation.

In order to fully exploit the benefits provided by OFDM modulation, large envelope variations before the RF portion of an OFDM transmitter must be avoided. Signal peaks can lead to saturation in the power amplifier (PA), which in turn increases out-of-band radiation, creates in-band distortion, and reduces PA efficiency. The PA dominates the power consumption of the communication system. Thus this decrease in efficiency results in lower battery life in mobile (wireless) devices and the need for sophisticated heat dissipation techniques in base stations. To deal with this important issue, advanced signal processing techniques are required, which have low implementation complexity.

Numerous techniques have appeared in the literature to reduce the PAPR [1]. They can largely be classified as distortion or distortionless techniques. Distortion techniques were introduced in [2–10]. They create undesirable in-band distortion [2], peak regrowth [4], or out-of-band radiation [5–9]. Many distortionless techniques have also been proposed [11–19]. They can reduce the PAPR without generating out-of-band radiation and/or in-band distortion. The effectiveness of distortionless techniques can be characterized in terms of the PAPR reduction, implementation complexity, and data rate loss. Schemes based on error correcting codes [11–15] sacrifice the data rate to improve PAPR performance. They require memory to store the codewords and introduce delay due to the time required to find low PAPR codewords, particularly when the number of subcarriers is large. Another class of distortionless techniques employ constellation mapping [16–18]. The constellation expansion in [16] requires a complex optimization process, particularly with a large number of subcarriers. Simpler constellation mapping techniques, which can be practically implemented, are active constellation extension [17] and tone reservation [18].

As distortionless phase optimization techniques, partial transmit sequence (PTS) [19], and selective mapping (SLM) [20] can provide significant PAPR reduction with a small amount of redundancy. With SLM, multiple sequences are generated by phase rotating the original data block and the sequence with the lowest PAPR is selected for transmission. Randomly chosen phase sequences lead to lower PAPR compared to other schemes such as sequences with adjacent or equally spaced subcarriers. In the PTS approach [19], disjoint subblocks of OFDM subcarriers are phase shifted separately after the IFFT is computed. If the subblocks are optimally phase shifted, they exhibit minimum PAPR and consequently reduce the PAPR of the merged signal. The number of subblocks and the corresponding partitioning determine the PAPR reduction. The search for optimum subblock phase factors is computationally complex, but this can be reduced with sphere decoding [21]. Pseudorandom (such as m-sequence), subblock partitioning has been found to provide better PAPR reduction compared to contiguous partitioning schemes. Typically, the receiver requires side information corresponding to the optimal phases in PTS and the transmitted sequences in SLM. Techniques for avoiding explicit side information transmission are presented in [22, 23]. SLM and PTS are known as average power preserving techniques as they do not increase the average signal power. The main drawback of PTS and SLM arises from the computation of multiple IFFTs, resulting in a complexity proportional to the number of PTS subblocks or SLM sequences.

Another class of distortionless techniques, tone reservation (TR) [16], increases the average power. With TR, a predefined set of OFDM subcarriers is reserved to generate peak reduction signals. The transmitter does not send data on these subcarriers, so they are orthogonal to the data subcarriers. Consequently, these added signals do not distort the data subcarriers, and recovering the data at the receiver is trivial. These peak reduction signals are used to compensate for high peaks in the signals on the data subcarriers. TR is particularly appropriate when there are a large number of subcarriers. The optimal choice of values for the reserved tones can be formulated as a quadratically constrained quadratic program for complex multicarrier signals (passband) and a linear program for real multicarrier signals (baseband) [16].

Suboptimal iterative algorithms such as gradient and controlled clipping [16] have been proposed, which have less complexity but slightly inferior PAPR performance compared to the optimal solutions. Hence, they provide a tradeoff between computational complexity and PAPR reduction. These algorithms, however, suffer from slow convergence after the first few iterations. In [24], the controlled clipper algorithm was considered. Improved performance was obtained by using filtered clipping noise as the peak-reduction signal. This noise was adaptively scaled to reduce the PAPR. However, the resulting complexity is still high since multiple FFT and IFFT operations are required during the iterative process.

In the suboptimal iterative algorithms, the PAPR reduction capability greatly depends on the location and number of peak reduction tones (PRTs). The locations also affect the convergence rate of the algorithm. As indicated in [16], the optimal peak reduction kernels can be computed off-line or during initialization if the channel is static. However, when the channel is not static, the PRT locations and consequently the peak reduction kernels should be updated according to the rate at which the channel changes [16]. This can create significant computational complexity. Side information must also be transmitted in order to identify the reserved tones at the receiver.

This paper first provides an overview of the major distortionless PAPR reduction techniques, namely, partial transmit sequence (PTS), selective mapping (SLM), and tone reservation (TR). Their main features are described and analyzed. We consider practical solutions for improving both the computational complexity and PAPR performance of these techniques. In doing so, we provide a review of the low-complexity techniques proposed in [25–28] that exploit the structure of the IFFT/FFT algorithms. These new techniques are based on the concept of identical inverse discrete Fourier transforms (IDFTs). The structure and properties of these identical IDFTs are used to improve both the PAPR and complexity.

The remainder of this paper is organized as follows. The structure of an OFDM transceiver is described in Section 2. We also characterize the PAPR, including a statistical description, and examine its effect on performance. Distortionless PAPR reduction techniques are introduced in Section 3. Section 4 presents the IFFT-based PAPR reduction techniques, and their performance and computational complexity are examined. Finally, Section 5 provides some concluding remarks and suggestions for future work.

We use the following notation. Uppercase and lowercase bold letters represent matrices and vectors, respectively. We use $\| \cdot \|_\infty$ to denote infinity-norm, $\| \cdot \|_2$ for 2-norm, $E[\cdot]$ for expectation, $(\cdot)^T$ for transpose, and $(\cdot)^*$ for complex conjugate.

OFDM SIGNALS AND PAPR

This section presents the OFDM system model and a characterization of the major disadvantage of OFDM, namely, a high peak-to-average power ratio (PAPR). We first introduce the OFDM transceiver structure and then describe the nonlinear power amplifier model used at the transmitter. Next, the effect of high peaks on the OFDM signal envelope is discussed.

The OFDM Transceiver

A block diagram of the OFDM transceiver is given in Figure 1. The serial input bit stream is sent to a quadrature-amplitude modulation (QAM) or phase-shift keying (PSK) constellation mapper, which outputs N parallel constellation points $X(k)$ representing the data. The vector $\mathbf{X}=[X(0),\dots,X(k),\dots,X(N-1)]^T$ denotes the frequency domain signal, which is a vector of QAM/QPSK complex symbols. The OFDM discrete time symbols $\mathbf{x}=[x(0),\dots,x(n),\dots,x(N-1)]^T$ are obtained by performing an IDFT on **x**. The parallel time domain samples $x(n)$ are then converted to a serial stream. A cyclic prefix is appended before the OFDM symbol **x**. This prefix should have a duration longer than the maximum delay due to the propagation paths [16]. This prevents intersymbol interference and enables simple single-tap equalization. The sequence is converted to an analog signal $x(t)$, up converted to the carrier frequency, amplified to obtain $\hat{x}(t)$, and the resulting signal is transmitted through the channel. At the receiver, the reverse operations are performed.

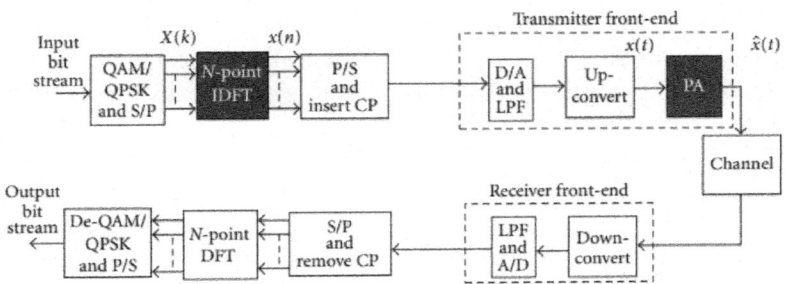

Figure 1: Block diagram of an OFDM transceiver with the power amplifier (PA).

The complex envelope of the baseband OFDM signal, defined over the time interval

$t \in [0, T_s]$ where T_s is the OFDM symbol duration, can be expressed as

$$x(t) = \frac{1}{N} \sum_{k=0}^{N-1} X(k) e^{j2\pi kt/T_s}.$$

(1)

The corresponding discrete time signal is

$$x(n) = \frac{1}{N} \sum_{k=0}^{N-1} X(k) e^{j2\pi kn/JN}, \quad n = 0, \ldots, N-1.$$

(2)

To simplify the notation, the time domain OFDM samples are represented as

$$\mathbf{x} = \mathbf{QX},$$

,(3)

where **Q** is an N-point IDFT matrix with elements

$$Q_{\alpha,n} \;=\; (1/N)e^{-j2\pi(\alpha+1)(n+\hat{1})/N}, 0 \leq \alpha \leq N-1, 0 \leq n \leq N-1.$$

In this paper, we consider a solid state power amplifier (SSPA) model [16] with amplitude modulation (AM) characteristic

$$\hat{x}(t) = \frac{hx(t)}{\left[1 + (hx(t)/x_{SAT})^{2\gamma}\right]^{1/2\gamma}},$$

$$(4)$$

where $\hat{x}(t)$ is the amplified OFDM signal and x_{SAT} is the output saturation level. The amplifier saturation power, P_{SAT}, is defined as $P_{SAT}=x^2{}_{SAT}$. The parameter γ controls the smoothness and h is the amplifier small signal gain. The AM/phase modulation (PM) conversion of the SSPA is assumed to be zero. In order to reduce the nonlinear distortion due to signal peaks, the amplifier is driven with an input back-off

$$IBO_{dB} = 10\log_{10}\frac{P_{SAT}}{E\left[|x(t)|^2\right]}.$$

$$(5)$$

Peak Power and Its Effects

If the modulated QAM/QPSK complex symbols $X(k)$ add constructively, they can generate a time domain signal with a large amplitude. Thus, the output signal $x(n)$ can have high peak values. Signal peaks much higher than the average can exceed the linear range of the amplifier, causing distortion of the OFDM signal. This distortion has an impact on the received signal constellation similar to additive noise if it occurs frequently [16]. Figure 2 illustrates the effect of the distortion for $N=256$ and a 16-QAM constellation. The in-band distortion is similar to additive Gaussian noise, which increases the bit error rate (BER) at the receiver. The out-of-band radiation (outside the spectrum of the OFDM signal) introduces interference in adjacent channels. Figure 3 shows the power spectral density (PSD) of an OFDM signal with $N=256$ subcarriers for different values of IBO. The spectral spreading is due to the nonlinear distortion.

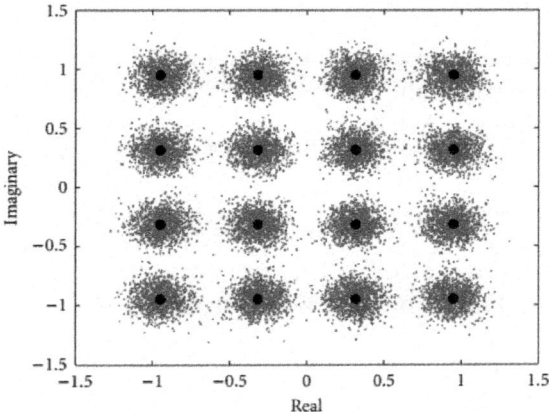

Figure 2: Effect of clipping noise on the received OFDM signal for 1000 OFDM symbols with a 16-QAM constellation, N=256, IBO=5 dB.

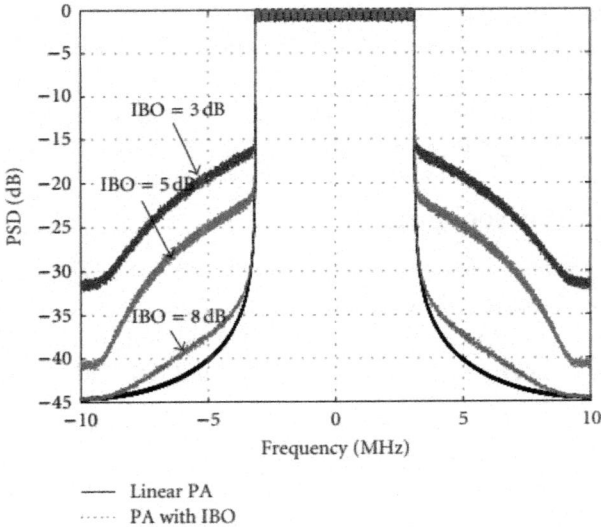

Figure 3: Power spectral density of an OFDM signal for different IBO values with $N = 256, J = 4, h = 1, x_{SAT} = 1$, and $\gamma = 10$.

To control the effects of nonlinear distortion, one can force (scale) the PA to operate in the linear region or increase the IBO. However, PAs are more power efficient when operating close to the saturation region. In addition, for the same transmit power, a larger IBO increases power consumption and the cost of hardware devices (e.g., a power amplifier with a large linear range).

Peak-to-Average Power Ratio (PAPR)

The most popular metric to evaluate the variation in the time domain signal is the peak-to-power average ratio (PAPR). For the OFDM system in Figure 1, the PAPR of $\mathbf{x}$ is defined asPAPR

$$\text{PAPR}(\mathbf{x}) = \frac{\|\mathbf{x}\|_\infty^2}{E\left[\|\mathbf{x}\|_2^2\right]/JN},$$

(6)

where J is the oversampling factor [16]. In order to evaluate the PAPR reduction, we employ the complementary cumulative distribution function

$$\text{CCDF} = \Pr[\text{PAPR}(\mathbf{x}) > \delta],$$

(7)

which represents the probability that the PAPR of a symbol exceeds the clipping level δ.

DISTORTIONLESS PAPR REDUCTION TECHNIQUES

In this section, we provide an overview of the most popular distortionless PAPR reduction techniques, PTS, SLM, and TR. They are based on data or signal modification prior to the power amplifier and reduce the PAPR without distorting the signal (which creates out-of-band radiation and/or in-band distortion). These techniques are described and analyzed below.

Partial Transmit Sequence

Partial Transmit Sequence (PTS) was introduced in [19] as a PAPR reduction scheme. This is achieved by generating different versions of the same symbol via phase-rotation of the time domain OFDM signal. With PTS (see Figure 4), the frequency domain vector $X(k)$ is partitioned into P disjoint subblocks

$$\mathbf{X}p = [X_p(0), \ldots, X_p(N-1)]^T, 0 \le p \le P-1$$

that are zero-padded to length N giving

$$\mathbf{X} = \sum_{p=0}^{P-1} \mathbf{X}_p.$$

.(8)

The combination of these subblocks with the phase rotation vectors

$$\mathbf{\Theta}_p = [\theta_p(1), \ldots, \theta_p(N-1)]$$

yields the alternative frequency domain vectors with

$$\mathbf{X}' = \sum_{p=0}^{P-1} \mathbf{\Theta}_p \mathbf{X}_p.$$

.(9)

The P subblocks are optimally phase rotated to achieve a reduced PAPR signal **x**. Since each subblock is independently rotated by a phase vector Θp, the phase vector multiplication can be performed after the IDFT computation. This is an advantage as the PAPR can be computed without changing between the time and frequency domains. Hence, we can take the IDFT of (9) and exploit the linearity of the IDFT to obtain

$$x' = \sum_{p=0}^{P-1} \Theta_p Q X_p = \sum_{p=0}^{P-1} \Theta_p x_p,$$

(10)

where the $x_p = QX_p$ are the P time domain partial transmit sequences.

The sequence x' with the smallest PAPR is chosen for transmission based on the following criterion

$$\left[\Theta_0, \ldots, \Theta_{P-1} \right] = \underset{\Theta_1, \ldots, \Theta_{P-1}}{\operatorname{argmin}} \left\{ \| x' \|_\infty^2 \right\}.$$

(11)

Assuming W is the number of phase values and $\Theta_0 = [1, \ldots, 1]^T$, the search complexity to find the lowest PAPR sequence is W^{P-1}. To improve efficiency, the phase factors can be restricted to values in the set $\theta_p(n) \in \{0, \pm 1, \pm j\}$. The search for the optimum subblock phase factors is computationally intensive, but this complexity can be reduced with sphere decoding [21]. PTS requires that $(P-1)\log_2 W$ bits per OFDM symbol be transmitted as explicit side information. This information is used at the receiver to recover the original data. Techniques to avoid side information are given in [23, 29].

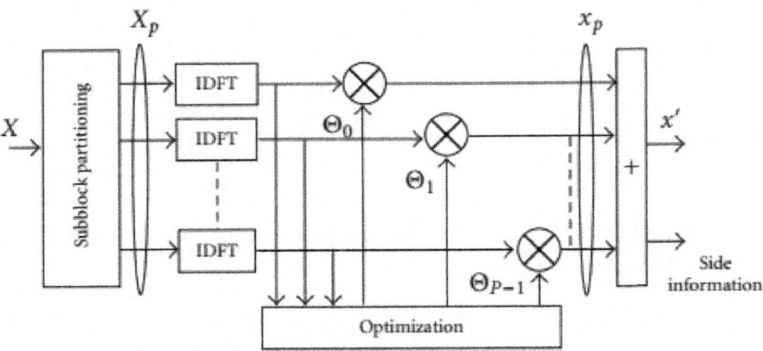

Figure 4: Block diagram for the partial transmit sequence PAPR reduction technique.

As shown in [19], pseudorandom subblock partitioning was found to have the best PAPR reduction compared to contiguous and other noncontiguous partitioning schemes. The autocorrelation function (ACF) of the PTS subblocks shows that this approach provides less correlated adjacent time samples. PTS is very effective at reducing the PAPR; however, the PAPR performance depends on W, P, and the method of subblock partitioning. If the number of subblocks, P, and/or the number of phase values, W, are increased, the PAPR reduction capability is improved. This is shown in Figure 5 for $N=256$.

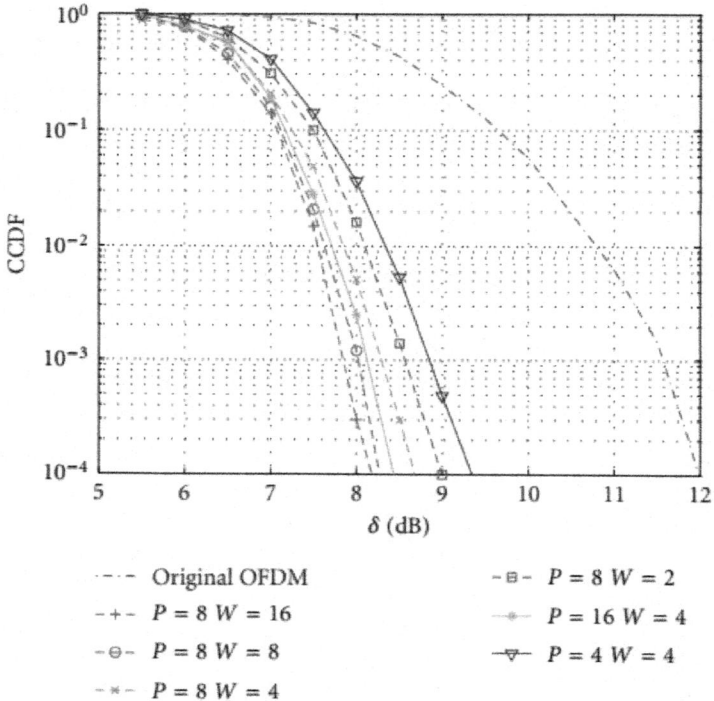

Line	Legend
⋯⋯	Original OFDM
-+-	$P = 8\ W = 16$
-⊖-	$P = 8\ W = 8$
-*-	$P = 8\ W = 4$
-⊟-	$P = 8\ W = 2$
-*-	$P = 16\ W = 4$
-▽-	$P = 4\ W = 4$

Figure 5: Complementary CDF of the OFDM signal for arious P and W with $N=256$and pseudorandom subblocks.

According to (10), the number of IDFT calculations that have to be computed is P, which is typically in the range from 2 to 16. Thus the resulting computational complexity can be high, particularly when N is large. IFFT-based PTS focuses on the computational complexity of the multiple IFFTs. This will be discussed in Section 4.

Selective Mapping

A simple approach to generate different mappings for the same OFDM symbol is to phase-rotate the frequency domain signal. This technique, called selective/selected mapping (SLM), was proposed in [30, 31]. The SLM concept is similar to that of PTS, but SLM phase rotates each subcarrier individually, while blocks of subcarriers are rotated in PTS. With SLM, multiple sequences are generated by multiplying independent phase sequences with the original data, and the sequence with the lowest PAPR is chosen for transmission (see Figure6). Consider a phase rotated version of $\mathbf{X}$ given by

$$\mathbf{U}_\omega = \mathbf{X} \cdot \mathbf{\Phi}_\omega, \tag{12}$$

where·denotes elementwise multiplication and $\mathbf{\Phi}_\omega = [\phi_w(1), \ldots, \phi_w(N-1)]$ represents the phase sequence vector. The time domain OFDM signal $\mathbf{u}\omega$ is obtained using the IDFT of $\mathbf{U}_\omega$. Hence, all of the candidate symbols carry the same information $\mathbf{X}$. In SLM, the lowest PAPR signal $\mathbf{u}_\omega'$ is chosen for transmission from theΩ candidate signals, including the original signal $\mathbf{x}$, that is

$$\mathbf{u}_{\omega'} = \arg \min_{1 \leq \omega \leq \Omega} \text{PAPR}[\mathbf{u}_\omega]. \tag{13}$$

It has been determined that using randomly chosen phase sequences provides better PAPR reduction than other sequences such as complementary Golay and Walsh-Hadamard sequences [32].

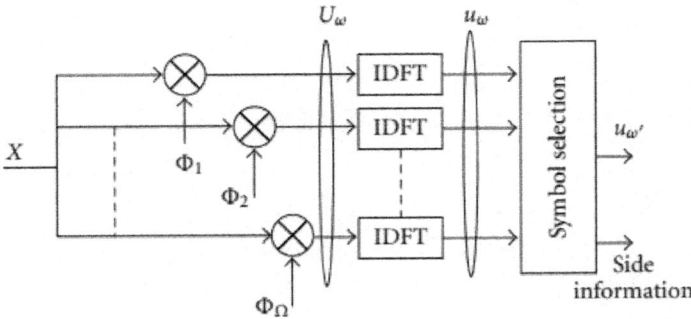

Figure 6: Block diagram for the selective mapping PAPR reduction technique.

Figure 7 presents the CCDF versus PAPR for SLM for various numbers of randomly chosen phase sequence vectors $\mathbf{\Phi}_\omega$ from the set $\{0, \pi\}$ with $N=256$ and $J=4$. This shows that larger PAPR reductions can be achieved as the number of SLM sequences Ω is increased. In addition, similar to

PTS, SLM requires Ω IDFTs to obtain the sequences $\mathbf{u}_\omega$. As a consequence, this techniques has significant computational complexity for typical values of Ω, between 2 and 16. Complexity reductions can be achieved by implementing the IFFT-based SLM technique described in Section 4.

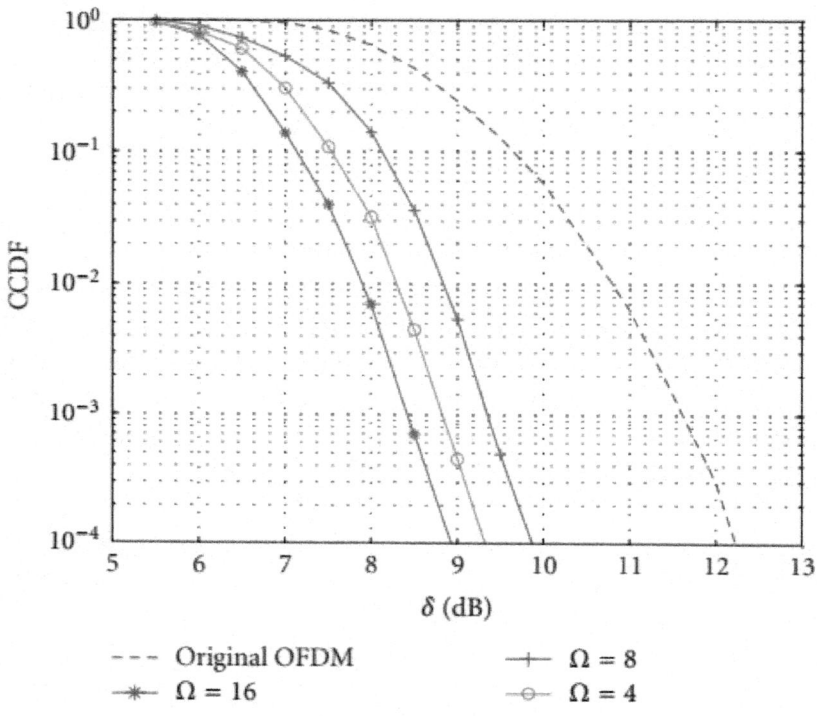

Figure 7: PAPR CCDF of SLM for various Ω with N=256 and J=4.

Tone Reservation

With the tone reservation (TR) technique, L subcarriers called peak reduction tones (PRTs) are reserved to generate a peak-reduction signal.

Let $\mathcal{R}=[\lambda_0,\ldots,\lambda_{L-1}]$ denote the ordered set of reserved tone positions, and $\mathcal{R}_c$ denote the complement of $\mathcal{R}$ in $\mathcal{N}=[0,\ldots,N-1]$. The frequency domain signal including PRTs can be expressed as

$$X(k) + C(k) = \begin{cases} C(k), & k \in \mathcal{R}, \\ X(k), & k \in \mathcal{R}^C, \end{cases}$$

$$(14)$$

where $C(k)$ is the PAPR reduction signal. The corresponding signal $\mathbf{c}$ is given by $\mathbf{c}=\mathbf{Q}\mathbf{C}$ where $\mathbf{C}=[C(0),...,C(N-1)]^T$. The PAPR after adding the peak-reduction signal $\mathbf{c}$ isPAPR

$$\text{PAPR}(\mathbf{x}+\mathbf{c}) = \frac{\|\mathbf{x}+\mathbf{c}\|_\infty^2}{E\left[\|\mathbf{x}\|_2^2\right]/JN},$$

(15)

where J is the oversampling factor. Since the optimal tone reservation solution, $\mathbf{c}$opt, only slightly increases the mean power, the denominator of (15) is not a significant function of $\mathbf{c}$ and so can be ignored [16]. This simplifies the calculation of $\mathbf{c}$opt. Hence, the goal is to minimize the peak power of the signal $\mathbf{x}+\mathbf{c}$. The optimal TR peak reduction signal $\mathbf{c}$ can then be formulated as the solution to the optimization problem [16]

$$\min_\mathbf{c}\|\mathbf{x}+\mathbf{c}\|_\infty = \min_\mathbf{C}\|\mathbf{x}+\mathbf{Q}\mathbf{C}\|_\infty,$$

(16)

where $\mathbf{C}$ is the corresponding frequency domain vector of $\mathbf{c}$.

To obtain a low-complexity solution, suboptimal iterative algorithms such as gradient and controlled clipper [16] have been developed to generate the peak reduction signal $\mathbf{c}$. They provide a trade-off between computational complexity and PAPR reduction. The gradient algorithm iteratively compute

$$\breve{\mathbf{x}}^{\beta+1} = \breve{\mathbf{x}}^\beta - \mu \sum_{|\breve{x}^\beta(n)|>x_{SAT}} a^\beta(n)\mathbf{e},$$

(17)

where

$a^\beta(n) = \breve{x}^\beta(n) - x_{SAT}e^{\angle(\breve{x}^\beta(n))}$ is a complex scalar at the βth iteration and $\mathbf{e}=\mathbf{Q}\mathbf{E}$. $\mathbf{E}=[E(0),...,E(N-1)]$ is called the frequency domain peak reduction kernel with binary elements $\{0,1\}$ according to the reserved subcarriers $\mathbf{C}$. A block diagram of the suboptimal iterative gradient algorithm is given in Figure 8. This algorithm first uses the PRT set to generate the kernel signal $\mathbf{e}$, then iteratively shifts this signal to the peak locations to reduce the amplitude of the OFDM signal. Figure 9 shows the performance of the gradient-based technique with $L/N=6.3\%$ reserved subcarriers chosen randomly, $\beta=40$ and $x_{SAT}=6$dB. In this case, the gradient algorithm achieves a PAPR reduction of 3 dB.

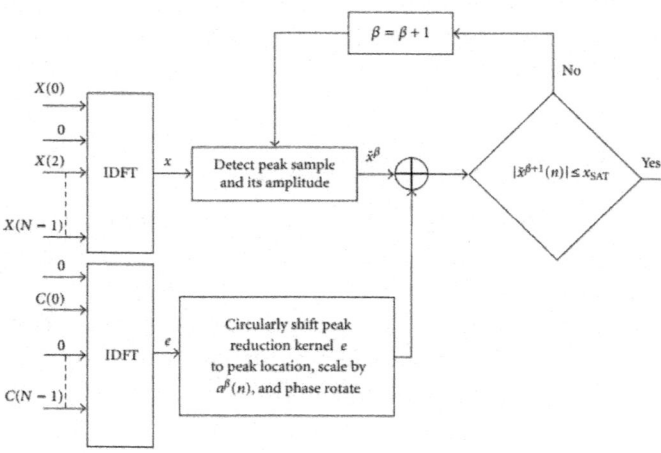

Figure 8: Block diagram for tone reservation PAPR reduction using the gradient algorithm.

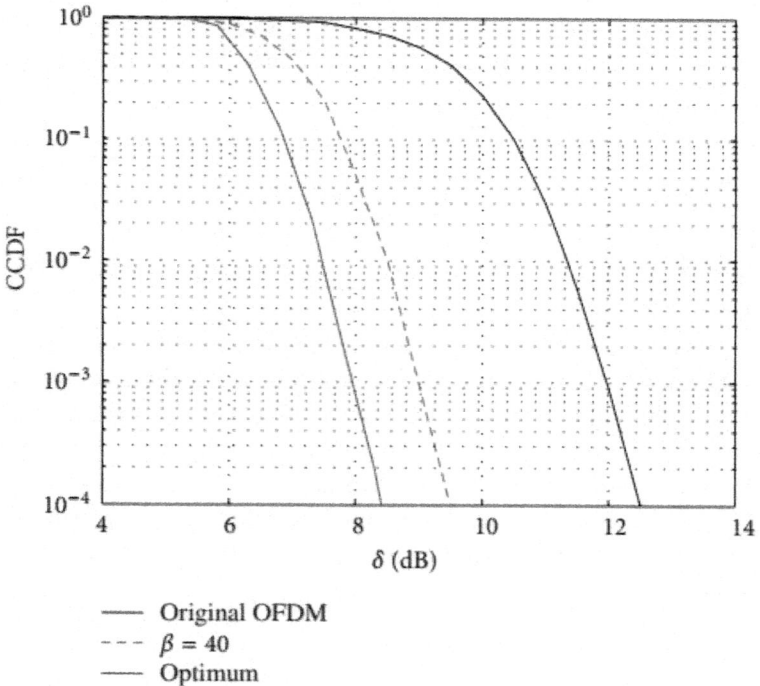

Figure 9: PAPR with the gradient-based algorithm in [16] using randomly chosen subcarriers for the PRTs, L/N=6.3%, β=40, and xSAT=6 dB.

The computational complexity of the gradient-based algorithm is essentially determined by the IFFT operations required to obtain the kernel signal **e**. If the channel is not static, the PRT locations should be updated periodically according to the coherence time of the channel. This can lead to very high computational complexity due to repeated calculations of the peak reduction kernels. Hence, we consider efficient computation of these kernels in Section 4.

IFFT-Based PAPR Reduction Techniques

This section presents a survey of IFFT-based PAPR reduction techniques [25–28]. It is shown that they have much lower complexity compared to the conventional distortionless methods given in the previous section. The performance of the IFFT-based PAPR reduction techniques and the previously developed solutions is also evaluated.

IFFT Algorithms

An IFFT algorithm converts the IDFT computation to $r \times N/r$-point DFTs iteratively through $m = \log_r N$ stages. As a consequence, the computational complexity is reduced from $\mathcal{O}(N^2)$ to $\mathcal{O}(N\log_r N)$. The value of r is called the radix. The PAPR reduction algorithms proposed in [25–28] exploit this recursive structure and the resulting identical IDFTs at each IFFT stage. This conversion also provides a means of analyzing and quantifying the effects of the intermediate signals in the transform in terms of PAPR reduction and computational complexity.

As described in [25], there are r^{v-1} identical N/r^{v-1}-point IDFTs at a particular radix-r IFFT stage v between stage v and the last stage m. Hence, from (3) the IFFT output corresponding to the inputs at a particular stage v can be expressed as

$$
\mathbf{x} = \begin{bmatrix} \mathbf{x}^1 \\ \vdots \\ \mathbf{x}^\eta \\ \vdots \\ \mathbf{x}^{r^{v-1}} \end{bmatrix} = \begin{bmatrix} \mathbf{Q}^1 \\ \vdots \\ \mathbf{Q}^\eta \\ \vdots \\ \mathbf{Q}^{r^{v-1}} \end{bmatrix} \begin{bmatrix} \mathbf{X}^1 \\ \vdots \\ \mathbf{X}^\eta \\ \vdots \\ \mathbf{X}^{r^{v-1}} \end{bmatrix}
$$

(18)

with v identical submatrices $\mathbf{Q}^1 = \mathbf{Q}^2 = \cdots = \mathbf{Q}^\eta = \cdots = \mathbf{Q}^{r^{v-1}}$. Figure 10 illustrates the IDFTs at stage v.

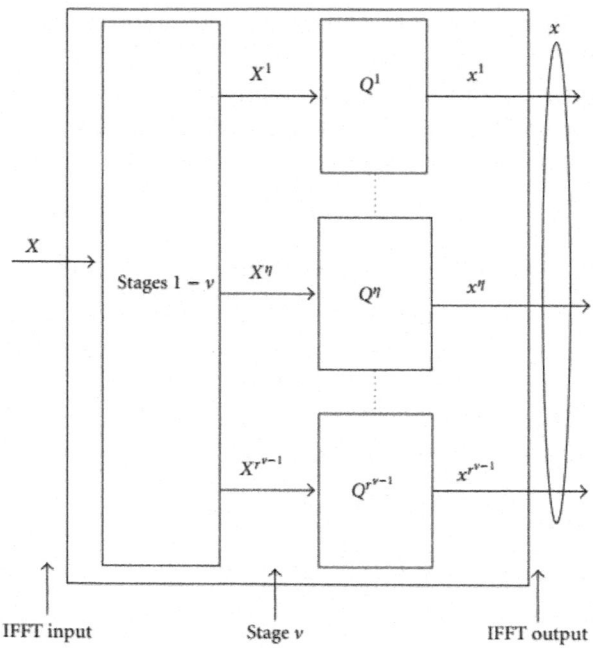

Figure 10: Block diagram of the identical IDFTs between intermediate stage v and the IFFT output.

Radix IFFT Computational Complexity

The radix-r IFFT algorithms derived above allow us to compute the computational complexity per stage. This consists of multiplicative and additive complexity. There are two major IFFT algorithms: decimation in time (DIT) and decimation in frequency (DIF) [25]. The multiplicative complexity of DIF algorithms; differs from that of DIT algorithms; however, both algorithms have the same additive complexity per stage.

From [25], the multiplicative complexity per stage for the radix-r DIF and DIT algorithms is defined as

$$B_v^{\text{mul,DIF}} = r^{v-1}\left(\frac{N}{r^v} - 1\right),$$

(19)

$$B_v^{\text{mul,DIT}} = r^{m-v}\left(\frac{N}{r^{m-v+1}} - 1\right),$$

(20)

Respectively.

The overall computational complexity for m stages is then given by [25]

$$B_{\text{total}}^{\text{mul,DIF}} = \sum_{v=1}^{m} B_v^{\text{mul,DIF}} = \sum_{v=1}^{m} r^{v-1}\left(\frac{N}{r^v} - 1\right),$$

(21)

The number of additions for the radix-r IFFT algorithm at stage v is

$$B_v^{\text{add}} = N(r - 1),$$

(22)

so the overall additive complexity for either algorithm is given by

$$B_{\text{total}}^{\text{add}} = vN(r - 1).$$

(23)

IFFT-Based PTS

As discussed previously, one of the major drawback of PTS arises from the computation of multiple IFFTs, resulting in a high complexity proportional to the number of subblocks. In order to reduce this complexity, the PTS technique proposed in [25] employs the input signals to the identical IDFTs at a given stage to obtain the PTS subblocks. This reduces the number of stages requiring multiple IFFTs.

DIF versus DIT Algorithms

The overall multiplicative complexity for IFFT-based PTS using DIF with P subblocks from (19) is

$$B^{\text{mul,DIF}} = \sum_{i=1}^{v-1} B_i^{\text{mul,DIF}} + P\sum_{i=v}^{m} B_i^{\text{mul,DIF}}.$$

.(24)

Similarly, from (20) the overall multiplicative complexity for IFFT-based PTS using DIT with P subblocks is

$$B^{\text{mul,DIT}} = \sum_{i=1}^{v-1} B_i^{\text{mul,DIT}} + P\sum_{i=v}^{m} B_i^{\text{mul,DIT}}.$$

(25)

From (23), both algorithms have the same additive complexity per stage, so the overall additive complexity for IFFT-based PTS with P subblocks is

$$B_{\text{add}} = Nr(r - 1)[v + P(m - v)].$$

(26)

The multiplicative complexity, which largely determines the computational complexity of the remaining stages, depends heavily on the type of IFFT algorithm. It was shown in [25] that DIT algorithms have a majority of the complex multiplication operations in the last stages, while DIF algorithms have a majority of these operations in the first stages. Thus, DIF has lower multiplicative complexity in generating the PTS subblocks compared to DIT, while both provide the same PAPR reduction. The PAPR performance was verified numerically in [25].

As an example, to compare the multiplicative complexity for $m-v$ remaining stages, consider N=256,2048,8192 for radix-2 DIF and DIT algorithms. This multiplicative complexity on a logarithmic scale is given in Figure 11.

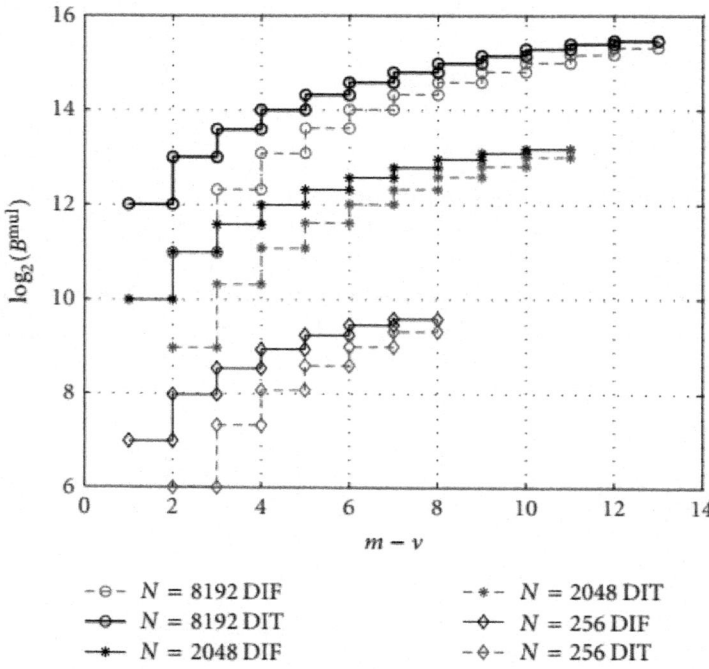

Figure 11: Number of nontrivial multiplications versus the number of remaining stages $m-v$ for DIF radix-2 compared with DIT radix-2 and N=256,2048,8192.

To evaluate the PAPR performance, the CCDFs of radix-2 DIF-PTS and DIT-PTS at stage $m-v$=5 and original PTS (O-PTS) are depicted in Figure 12 for P=8 and N=256,2048. This shows that the algorithms provide similar performance. In this case, DIF-PTS achieves a multiplicative complexity reduction ratio of 30%over DIT-PTS. This value increases to 43% and 58% when compared to O-PTS. It has been shown that high radix algorithms provide better PAPR reduction per stage compared to low radix algorithms and also have lower multiplicative complexity.

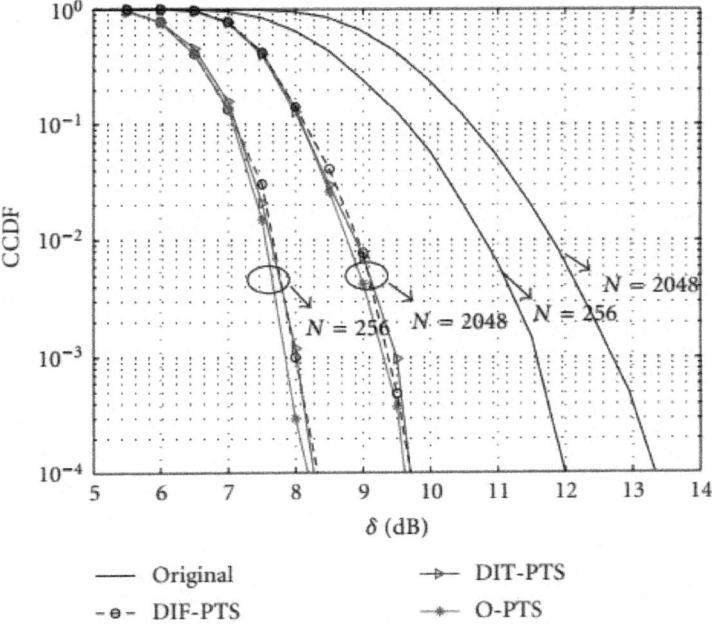

Figure 12: CCDF of DIF-PTS compared with DIT-PTS for $r=2$ at stage $m-v=5$ and O-PTS with $N=256,2048$.

Although IFFT-based PTS reduces the computational complexity, the PAPR reduction decreases as the number of stages after PTS partitioning is reduced. To achieve PAPR reduction close to that of O-PTS, there should be a sufficient number of stages remaining, so there is a limit on the achievable computational complexity reduction. Thus the major challenge is to decrease this complexity while maintaining a PAPR reduction close to that of O-PTS. To address this, the PTS technique in [27] was developed based on the normalized periodic auto-correlation function (ACF) of the PTS subblocks, which is defined as

$$F_p(\xi) = \frac{1}{\sigma_x^2} E\left\{ \left[x_p((n+\xi) \bmod N) \right] \left[x_p(n) \right]^* \right\},$$

(27)

where $\sigma_x^2 = E[|X(k)|]^2$. This represents the correlation between ξ-spaced complex samples in a subblock p. The ACF can be used in the design of the PTS subblocks to reduce both the PAPR and computational complexity. It was shown that the previously proposed pseudorandom subblock designs are not optimal for the case of IFFT-based PTS as they introduce repeated subcarriers (identical inputs to different IDFTs at a given stage are defined as repeated subcarriers) within a subblock. The effect of repeated subcarriers on the ACF of the subblocks is illustrated in Figure 13 where the number of subcarriers and subblocks is 32 and 4, respectively. The same pseudorandom sequence [01023312] was used for the inputs to the 4×8-point IDFTs using a radix-2 algorithm at Stage 3. This shows that repeated subcarriers result in a large ACF for the PTS sequences.

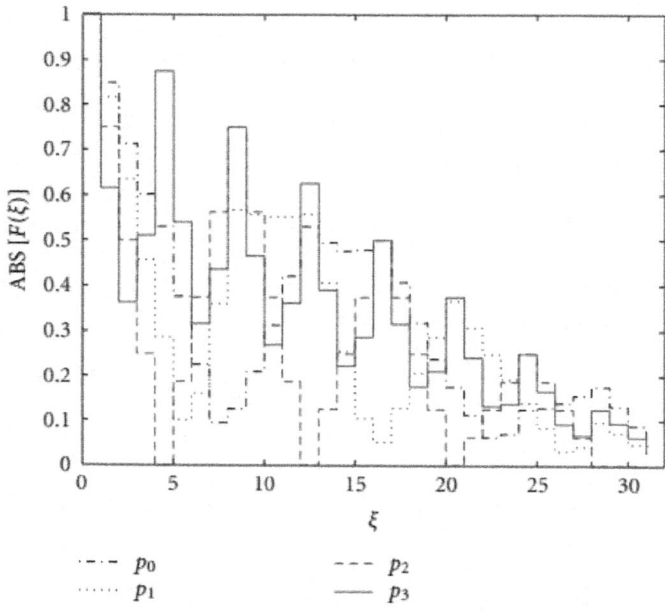

Figure 13: Autocorrelation of the PTS sequences for $N=32$, $v=3$, $P=4$, and $r=2$ with repeated subcarriers within each subblock.

PTS subblock partitioning was developed based on error-correcting codes (ECCs) to limit the number of repeated subcarriers [26]. This partitioning improves the ACF properties of the PTS subblocks as it provides less correlated adjacent time samples compared with other partitioning schemes. In fact, the proposed ECC technique minimizes the number of repeated subcarriers within the subblocks at a particular stage v. The ECCs are repetition codes (RCs) over ZP, the integer ring of P elements. As an example, consider the 4×8-point IDFTs using radix-2 at Stage 3. The m-sequence (MS) subblocks from [33] with $P=4$ are

$$
\mathcal{S}_{MS} = \begin{bmatrix} 0 & 2 & 3 & 1 & 2 & 1 & 0 & 2 \\ 1 & 0 & 0 & 0 & 2 & 1 & 2 & 1 \\ 2 & 3 & 3 & 1 & 2 & 3 & 1 & 0 \\ 0 & 2 & 3 & 3 & 3 & 3 & 1 & 0 \end{bmatrix}.
$$

$$.(28)$$

Each row in (28) represents the input to an 8-point IDFT. The repeated subcarriers within each subblock can be identified from the repeated numbers in the columns of the matrix. This shows that there are as many as three repeated subcarriers within the subblocks. The subblocks obtained using ECCs over $\mathcal{Z}_4$ are

$$\mathcal{S}_{\text{ECC}} = \begin{bmatrix} 0 & 1 & 0 & 2 & 3 & 3 & 1 & 2 \\ 1 & 2 & 1 & 3 & 0 & 0 & 2 & 3 \\ 2 & 3 & 2 & 0 & 1 & 1 & 3 & 0 \\ 3 & 0 & 3 & 1 & 2 & 2 & 0 & 1 \end{bmatrix}. \tag{29}$$

Note that in this case there are no repeated subcarriers within the subblocks. Figures 14 and 15 present the absolute value of the ACF vectors for subblocks $p0$ to $p3$ with MS and ECC subblocks, respectively. The ECC subblocks show a significant reduction in the ACF compared with the MS subblocks. In fact, the ECC subblocks provide an ACF, which is nearly flat.

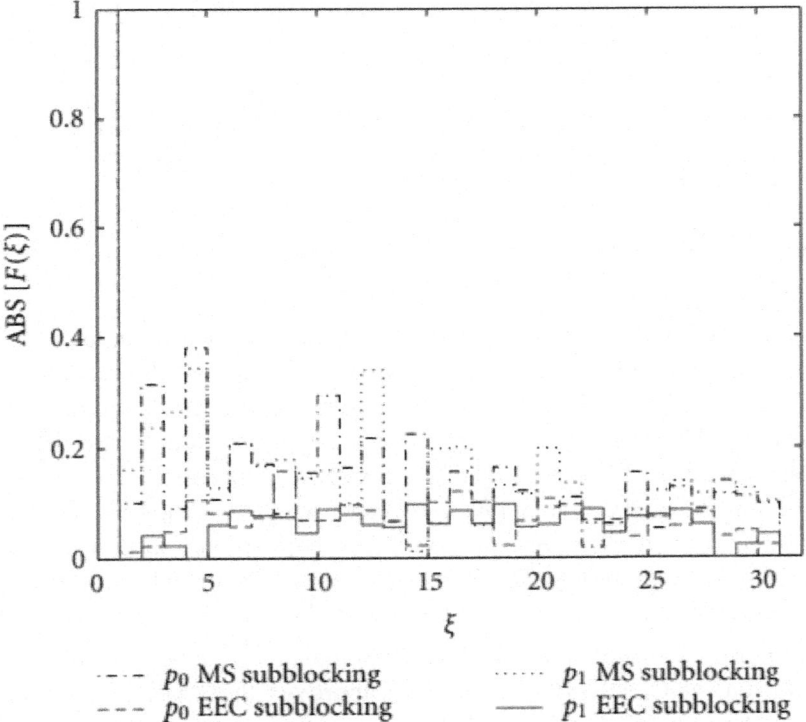

- -·- p_0 MS subblocking   ······ p_1 MS subblocking
- - - p_0 EEC subblocking   —— p_1 EEC subblocking

Figure 14: Autocorrelation of the PTS sequences for N=32, v=3, P=4, r=2, with m-sequence and ECC subblocks, p_0 and p_1.

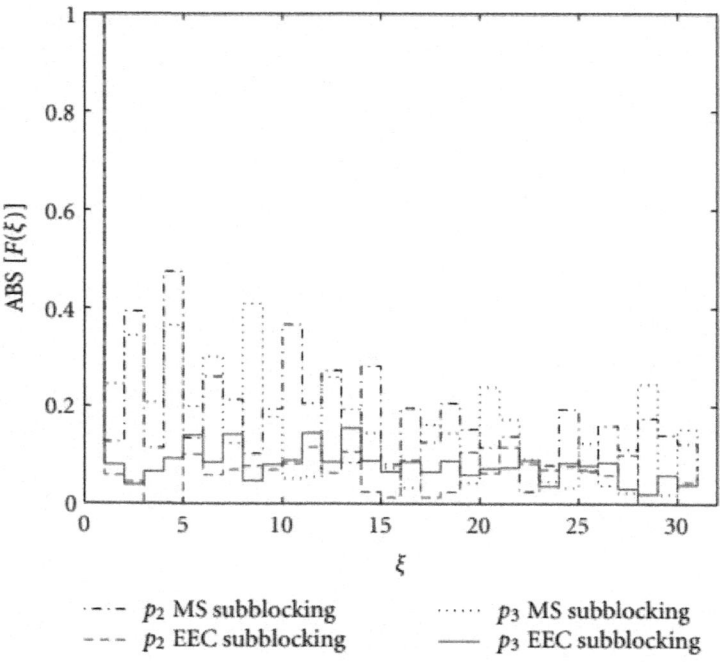

Figure 15: Autocorrelation of the PTS sequences for $N=32$, $v=3$, $P=4$, $r=2$, with m-sequence and ECC subblocks, $p2$ and $p3$.

As stated in [26], the IFFT and FFT operations use the same coefficients. Hence, the FFT coefficients used to recover the data at the receiver are the same as the IFFT coefficients used to generate the OFDM symbols at the transmitter. However, we must take into account the order of the IFFT inputs at the transmitter. If we assume these inputs are in normal order, the inputs to the FFT at the receiver should be in reverse order. Thus, the FFT computations at the receiver are symmetric to the IFFT computations, so if stage v is used to obtain the PTS subblocks, the data is recovered at FFT stage $m-v$ at the receiver. Hence, the side information required is the same as with O-PTS.

Figure 16 presents the PAPR performance for $N=256$ and $P=8$. This shows that at CCDF=10^{-4}, PTS-ECC improves the PAPR performance by approximately 2 dB for 2 and 3 remaining stages compared to PTS-MS. The corresponding computational complexity reduction over O-PTS is from 45% to 52% [25].

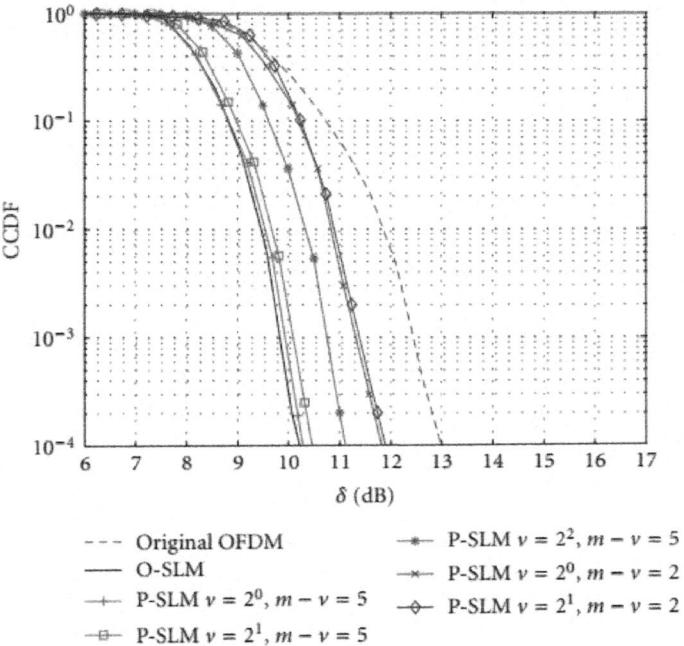

Figure 16: PAPR CCDF for PTS-ECC with 2 and 3 remaining stages, PTS-MS and O-PTS for N=256 and P=8.

IFFT-Based SLM

Similar to PTS, a major limitation with SLM is the computation of multiple IFFTs, resulting in a high complexity for practical systems, which is proportional to the number of SLM sequences. This complexity can be reduced with IFFT-based SLM [26], which uses the product of intermediate signals within the IFFT and the SLM phase sequences. Multiple IFFTs are then applied only to the remaining stages.

To reduce the computational complexity of the remaining stages, DIF is employed with IFFT-based SLM rather than DIT, as the former results in fewer multiplication. In addition (similar to PTS), IFFT-based SLM using a higher radix FFT algorithm results in lower multiplicative complexity than using a lower radix. To further reduce the computational complexity of IFFT-based SLM, partial SLM was proposed in [26]. In this case, multiple IFFTs are computed over a fraction of the identical IDFTs for the remaining stages, so that only a subset of the inputs **X**η are phase rotated to obtain the time domain sequences **U**w. It was shown that partial SLM significantly lowers the computational complexity compared to original SLM (O-SLM). As stated in [26], a complexity reduction of 75% over O-SLM can be obtained with 5 remaining stages and 8 SLM sequences. Figure 17 presents the CCDF of O-SLM and partial SLM for various numbers of remaining stages and subsets v. This shows that partial SLM with v=2 provides PAPR performance close to that with O-SLM. Therefore, this technique provides a better PAPR versus complexity tradeoff compared to O-SLM.

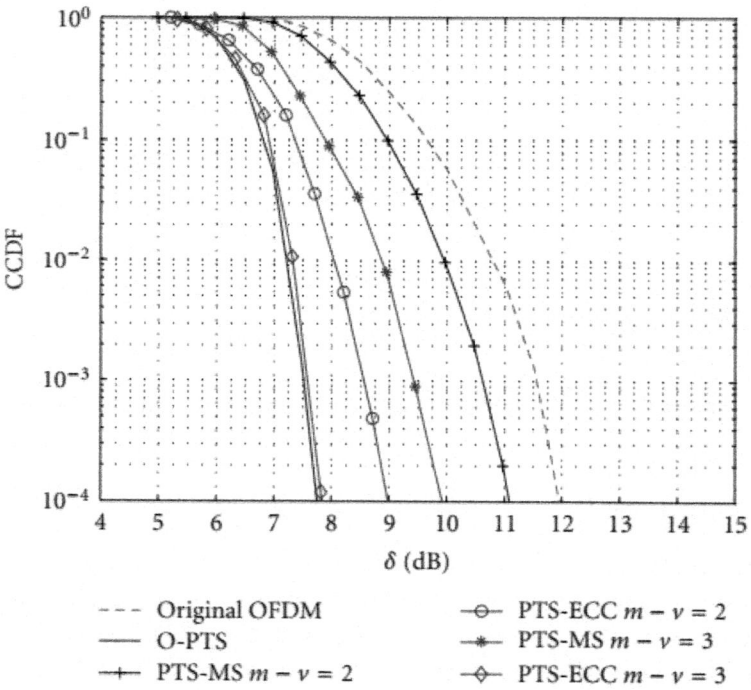

Figure 17: PAPR CCDF for O-SLM and partial SLM for various numbers of remaining stages $m-v$ and subsets, v, and 8 SLM sequences with r=2 and N=2048.

To recover the data at the receiver, similar to IFFT-based PTS, only the IFFT input order at the transmitter must be taken into account. Therefore, the side information is the same as with original SLM.

IFFT-Based TR

As previously discussed, with the suboptimal iterative algorithms the PAPR reduction capability of TR depends strongly on the locations and number of peak reduction tones (PRTs). These locations also affect the convergence rate of the solution. As indicated in [16], the optimal peak reduction kernels can be computed off-line or during the initialization process if the channel is static. However, when the channel is not static, the PRT locations and consequently the peak reduction kernels should be updated as the channel varies [16]. This creates significant computational complexity. For example, to obtain a good PRT set (with a PAPR loss less than 0.1 dB from the optimal case), a complexity of one IFFT per set is reported in [16], with 11 to 28 randomly chosen sets of reserved subcarriers required. Side information must also be transmitted in order to identify the reserved tones at the receiver.

In [28], IFFT-based TR was proposed as an efficient gradient-based algorithm for PAPR reduction. This algorithm utilizes the FFT algorithms described previously to reduce the computational complexity associated with optimizing the peak reduction kernels. It was shown that this can significantly reduce the computational complexity in generating the kernels. Thus they can be updated efficiently when the channel is not static. IFFT-based TR has significantly less computational complexity compared with the approaches in [16]. With IFFT-based TR, the reserved tones are chosen over the inputs $\mathbf{X}\eta$ to the identical IDFTs at a given stage. In fact, a subset of PRT locations for each of the identical $N/^{rv-1}$-point IDFTs is used to generate the peak reduction kernels $\mathbf{e}\eta$. This results in a low-complexity solution to computing the kernels, which is appropriate when the PRTs must be updated frequently. The results presented in [28] demonstrate that the proposed algorithm has PAPR performance close to that of the gradient algorithm in [16], at a cost of one to two additional bits of side information.

The PAPR performance of IFFT-based TR and the algorithm in [16] are compared in Figure 18 for various numbers of iterations (β_p), with $N=512$, $r=2$, $v=4$, and $\beta=40$. This shows that the IFFT-based TR algorithm with 8 and 16 iterations outperforms the algorithm in [16] with 40 iterations. In this case, the IFFT-based TR technique also provides a complexity reduction of 56% [28]. The proposed IFFT-based TR algorithm thus provides a better tradeoff between complexity and PAPR performance compared to the gradient algorithm in [16], as these performance metrics are a function of r and v.

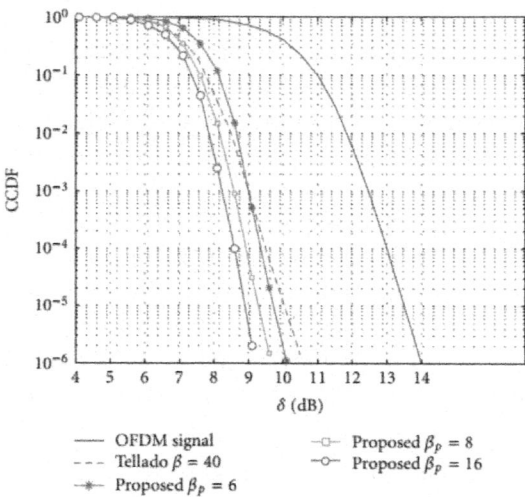

- —— OFDM signal
- --- Tellado $\beta = 40$
- —*— Proposed $\beta_p = 6$
- —□— Proposed $\beta_p = 8$
- —○— Proposed $\beta_p = 16$

Figure 18: PAPR CCDF for IFFT-based TR and the algorithm in [16] for various values of βp, with $L=32$ and $\beta=40$.

It should be noted that a radix-r IFFT algorithm can be practically implemented using radix-r butterflies with N/r butterflies per stage. Each stage of the IFFT operation reads N memory locations containing the butterfly inputs, processes the inputs, and writes them back. Since the proposed IFFT algorithm uses only $m-v$ of the m stages, the number of butterfly processing elements is reduced from mN/r to $(m-v)N/r$. In addition, since we compute the peak reduction kernel for only one N/r^{v-1}-point IDFT, the number of butterfly processing elements is only $(m-v)^{N/r^{v-1}}$. Using N/r^{v-1}-point IDFTs instead of N-point IDFTs significantly reduces the number of butterfly operations required.

CONCLUSIONS

A time domain OFDM signal can exhibit a large peak-to-average power ratio (PAPR). This reduces the efficiency of the power amplifier and causes nonlinear distortion, which increases out-of-band radiation and creates in-band distortion. One solution to this problem is to employ an expensive power amplifier with a large linear range. More practical techniques decrease the PAPR of the transmitted signal by modifying of the OFDM signal prior to the power amplifier.

There have been a variety of techniques developed to generate OFDM symbols with reduced PAPR, but none provide a large reduction in PAPR with low complexity and without degrading the performance of the system (distorting the OFDM signal). The most popular distortionless PAPR reduction techniques, partial transmit sequence (PTS), selective mapping (SLM), and tone reservation (TR) provide significant PAPR reduction. However, all have relatively high computational complexity. The tradeoff between PAPR reduction and computational complexity has motivated the development of numerous techniques.

IFFT-based PTS provides a solution to the major drawback of PTS, namely, the computational complexity due to multiple IFFTs. To generate the time domain PTS sequences, multiple transforms were computed over identical IDFTs. The periodic autocorrelation function (ACF) of the time domain PTS sequences was employed to develop a PTS subblock partitioning technique using error-correcting codes (ECCs). This minimizes the number of repeated subcarriers within a subblock and provides better PAPR reduction than pseudorandom subblocks. A PAPR reduction comparison between ECC subblocks and previous approaches was presented. This showed that the PTS subblocks using ECCs provides significant PAPR reduction with a small number of remaining stages. Hence, both PAPR and complexity reduction are achieved with this technique.

Similar to PTS, IFFT-based SLM can reduce the computational complexity due to the computation of multiple IFFTs. To generate the SLM sequences, intermediate signals within the IFFT were phase rotated. Further, a low-complexity SLM technique based on partial phase rotated inputs to the identical IDFTs was examined. This technique computes multiple inverse IFFTs with significantly lower computational complexity. A comparison between this technique and original SLM was presented in terms of PAPR reduction and computational complexity. This showed that the partial SLM approach provides significant complexity reduction with PAPR performance very close to that of original SLM.

A new class of gradient-based tone reservation algorithms, IFFT-based TR, was presented. To generate the peak reduction kernels, the transform matrices of identical IDFTs were employed. It was shown that they can be used to provide low-complexity solutions to determining the peak reduction tones (PRTs) and computing the peak reduction kernels. This is particularly important with time varying channels, in which case the algorithms must be employed periodically according to the coherence time of the channel. The cost is a slight increase in side information. Results were presented, which demonstrate that the IFFT-based TR technique outperforms the previous gradient algorithm in terms of computational complexity, with a slightly degradation in PAPR performance.

Future Work

Low-complexity IFFT algorithms are desirable for efficient transform implementation in hardware and software. The design of IFFT algorithms in the context of IFFT-based PAPR reduction presents new signal processing opportunities for multicarrier modulation. IFFT algorithms such as split-radix, radix-2/4/8, and radix-22 are interesting from the point of view of computational complexity, PAPR reduction, and suitability for hardware implementation. Thus this is an important research direction for the future. Optimization of the IFFT-based techniques introduced in this paper for implementation on digital signal processors (DSP), field programmable gate arrays (FPGAs), or application-specific integrated circuits (ASICs), should also be investigated.

Multiple-input multiple-output OFDM (MIMO-OFDM) systems and the related signal processing have acquired great significance in recent years. It is employed in fourth generation (4G) networks and has been proposed for broadband wireless communication systems. Thus, efficient MIMO-OFDM transmission and reception techniques are of interest. Similar to single antenna OFDM, one of the major drawbacks with MIMO-OFDM is that the signals transmitted on different antennas may exhibit a large PAPR. If one of the distortionless PAPR techniques is employed, complexity becomes a severe problem, much more so than in a single antenna OFDM system. This is because each transmitter implements the OFDM modulation. Previously developed PAPR reduction techniques for MIMO-OFDM do not address the computational issues associated with practical implementation. Thus, the development of efficient algorithms for IFFT-based PAPR reduction in space, frequency, or/and time block coded MIMO-OFDM systems is an important direction for future research.

REFERENCES

1.   S. H. Han and J. H. Lee, "An overview of peak-to-average power ratio reduction techniques for multicarrier transmission," IEEE Wireless Communications, vol. 12, no. 2, pp. 56–65, 2005.

2.   X. Li and L. J. Cimini, "Effects of clipping and filtering on the performance of OFDM," IEEE Communications Letters, vol. 2, no. 5, pp. 131–133, 1998. ·

3.   D. Wulich, N. Dinur, and A. Glinowiecki, "Level clipped high-order OFDM," IEEE Transactions on Communications, vol. 48, no. 6, pp. 928–930, 2000.

4.   H. Ochiai and H. Imai, "Performance analysis of deliberately clipped OFDM signals," IEEE Transactions on Communications, vol. 50, no. 1,

pp. 89–101, 2002.

5. M. Sharif, M. Gharavi-Alkhansari, and B. H. Khalaj, "On the peak-to-average power of OFDM signals based on oversampling," IEEE Transactions on Communications, vol. 51, no. 1, pp. 72–78, 2003.

6. R. Raich, H. Qian, and G. T. Zhou, "Optimization of SNDR for amplitude-limited nonlinearities," IEEE Transactions on Communications, vol. 53, no. 11, pp. 1964–1972, 2005.

7. Gatherer and M. Polley, "Controlling clipping probability in DMT transmission," in Proceedings of the 31st Asilomar Conference on Signals, Systems & Computers, pp. 578–584, November 1997.

8. J. Armstrong, "Peak-to-average power reduction for OFDM by repeated clipping and frequency domain filtering," Electronics Letters, vol. 38, no. 5, pp. 246–247, 2002.

9. X. Huang, J. Lu, J. Zheng, J. Chuang, and J. Gu, "Reduction of peak-to-average power ratio of OFDM signals with companding transform," Electronics Letters, vol. 37, no. 8, pp. 506–507, 2001.

10. X. Huang, J. Lu, J. Zheng, K. B. Letaief, and J. Gu, "Companding transform for reduction in peak-to-average power ratio of OFDM signals," IEEE Transactions on Wireless Communications, vol. 3, no. 6, pp. 2030–2039, 2004.

11. T. A. Wilkinson and A. E. Jones, "Minimisation of the peak to mean envelope power ratio of multicarrier transmission schemes by block coding," in Proceedings of the IEEE 45th Vehicular Technology Conference, pp. 825–829, July 1995.

12. J. A. Davis and J. Jedwab, "Peak-to-mean power control in OFDM, Golay complementary sequences, and Reed-Muller codes," IEEE Transactions on Information Theory, vol. 45, no. 7, pp. 2397–2417, 1999.

13. D. Wulich, "Reduction of peak to mean ratio of multicarrier modulation using cyclic coding,"Electronics Letters, vol. 32, no. 5, pp. 432–433, 1996. ·

14. V. Tarokh and H. Jafarkhani, "On the computation and reduction of the peak-to-average power ratio in multicarrier communications," IEEE Transactions on Communications, vol. 48, no. 1, pp. 37–44, 2000.

15. K. G. Paterson, "Generalized Reed-Muller codes and power control in OFDM modulation," IEEE Transactions on Information Theory, vol. 46, no. 1, pp. 104–120, 2000.

16. J. Tellado, Peak to average power reduction for multicarrier modulation [Ph.D. thesis], Stanford University, 2000.

17. B. S. Krongold and D. L. Jones, "PAR reduction in OFDM via active constellation extension," IEEE Transactions on Broadcasting, vol. 49, no. 3, pp. 258–268, 2003.

18. B. S. Krongold and D. L. Jones, "An active-set approach for OFDM PAR reduction via tone reservation,"IEEE Transactions on Signal Processing, vol. 52, no. 2, pp. 495–509, 2004.

19. S. H. Mueller and J. B. Huber, "Novel peak power reduction scheme for OFDM," in Proceedings of the International Symposium on Personal, Indoor and Mobile Radio Communications (PIMRC ‹97), pp. 1090–1094, September 1997.

20. M. Breiling, S. H. Müller-Weinfurtner, and J. B. Huber, "SLM peak-power reduction without explicit side information," IEEE Communications Letters, vol. 5, no. 6, pp. 239–241, 2001.

21. Alavi, C. Tellambura, and I. Fair, "PAPR reduction of OFDM signals using partial transmit sequence: an optimal approach using sphere decoding," IEEE Communications Letters, vol. 9, no. 11, pp. 982–984, 2005.

22. N. Carson and T. A. Gulliver, "Peak-to-average power ratio reduction of OFDM using repeat-accumulate codes and selective mapping," in Proceedings of the IEEE International Symposium on Information Theory, p. 244, July 2002.

23. D. S. Jayalath and C. Tellambura, "SLM and PTS peak-power reduction of OFDM signals without side information," IEEE Transactions on Wireless Communications, vol. 4, no. 5, pp. 2006–2013, 2005.

24. L. Wang and C. Tellambura, "An adaptive-scaling tone reservation algorithm for PAR reduction in OFDM systems," in Proceedings of the IEEE Global Telecommunications Conference (GLOBECOM ‹06), pp. 1–5, December 2006.

25. Ghassemi and T. A. Gulliver, "A low-complexity PTS-based radix FFT method for PAPR reduction in OFDM systems," IEEE Transactions on Signal Processing, vol. 56, no. 3, pp. 1161–1166, 2008.

26. Ghassemi and A. T. Gulliver, "Partial selective mapping OFDM with low complexity IFFTs," IEEE Communications Letters, vol. 12, no. 1, pp. 4–6, 2008.

27. Ghassemi and T. Gulliver, "PAPR reduction of OFDM using PTS and error-correcting code subblocking—transactions papers," IEEE Transactions on Wireless Communications, vol. 9, no. 3, pp. 980–989, 2010.

28. Ghassemi and T. A. Gulliver, "A simplified suboptimal algorithm for tone reservation OFDM," inProceedings of the IEEE International Conference on Communications (ICC ‹09), pp. 1–5, June 2009.

29. C. Feng, C. Y. Wang, C. Y. Lin, and Y. H. Hung, "Protection and transmission of side information for peak-to-average power ratio reduction of an OFDM signal using partial transmit sequences," inProceedings of the IEEE 58th Vehicular Technology Conference (VTC ‹03), pp. 2461–2465, October 2003.

30. R. W. Bäuml, R. F. H. Fischer, and J. B. Huber, "Reducing the peak-to-average power ratio of multicarrier modulation by selected mapping," Electronics Letters, vol. 32, no. 22, pp. 2056–2057, 1996.·

31. S. H. Muller and J. B. Huber, "Comparison of peak power reduction schemes for OFDM," in Proceedings of the IEEE Global Telecommunications Mini-Conference, pp. 1–5, November 1997.

32. N. Ohkubo and T. Ohtsuki, "Design criteria for phase sequences in selected mapping," in Proceedings of the 57th IEEE Semiannual Vehicular Technology Conference (VTC ‹03), pp. 373–377, April 2003.

33. W. Lim, S. J. Heo, J. S. No, and H. Chung, "A new PTS OFDM scheme with low complexity for PAPR reduction," IEEE Transactions on Broadcasting, vol. 52, no. 1, pp. 77–82, 2006.

Chapter 2

REAL-TIME COMMUNICATIONS IN AUTONOMIC NETWORKS: SYSTEM IMPLEMENTATION AND PERFORMANCE EVALUATION

C. Tselios[1] C. Papageorgiou[1] K. Birkos[1] I. Politis[1] and T. Dagiuklas[2]

[1]Wireless Telecommunications Laboratory, University of Patras, TK 26504, Patras, Greece

[2]Department of Telecommunications Systems and Networks, TEI of Messolonghi, TK 30300, Nafpaktos, Greece

ABSTRACT

This paper describes the design and prototype implementation of a communication platform aiming to provide voice and video communication in a distributed networking environment. Performance considerations and network characteristics have also been taken into account in order to provide the set of properties dictated by the sensitive nature and the real-time characteristics of the targeted application scenarios. The proposed system has been evaluated both by experimental means as well as subjective tests taken by an extensive number of users. The results show that the proposed platform operates seamlessly in two hops, while in the four hops scenario, audio and video are delivered with marginal distortion. The conducted survey indicates that the user experience in terms of Quality of Service has obtained higher scores in the scenario with the two hops.

INTRODUCTION

Mobile ad hoc networks have received particular attention the last years due to the wide range of applications such as real-time communications including video apart from voice where existing telecommunication infrastructure may fail. The introduction of low-cost wireless technologies and the standardization efforts of the IETF MANET Working Group have been generating renewed interest in research and development of MANETs outside the military field. The advent of new products in both hardware and software has eliminated many of the barriers of the past, enabling the development of integrated platforms providing a wide spectrum of services.

In this dynamic and distributed environment it is important to deploy multimedia application and services. This necessitates the deployment of P2P Voice and Video in a large scale, since many users become aware of the abilities of these newly developed architectures and migrate to them. Wireless multihop networks often show great potential, due to some characteristics such as node mobility and extended packet-forwarding ability [1]. Simply applying current peer-to-peer overlay techniques to MANETs is rather undesirable due to node mobility, energy consumption, and lack of infrastructure. Always keeping in mind that the overlay technology needs for power consumption and response time reasons to reflect the underlined physical network topology, wired network control schemes are unable to accommodate a constantly changing peer group where nodes constantly join and quit. An additional issue in the overall idea of this network architecture is peer cooperation. As shown in [2], most topology control algorithms assume that peers are cooperative, which is simply not the case. Peers are always trying to minimize their own costs such as the number of necessary communication links or the distance to other peers. Several studies [3] investigate the selfish peer impact on topology unfortunately in a rather theoretical approach where peers have global knowledge that is considered fundamental for overlay construction. Due to lack of a practical overlay topology control algorithm other means need to be established. Since traditional approaches tend to show decreased performance, peer-to-peer services in MANETs might need a fresh new design which would enable better results. It is certainly not a coincidence that even network operators are searching methods of using these novel applications in terms of profit [4]. Alas, the main issue in those topologies is no other than network capacity. A performance evaluation presented in [5] shows the influence of intra and interflow interference in channel utilization, which directly impacts the VoIP capacity. In more controlled environments such problems are not that obvious or additional components such as wireless mesh routers [6] could be an effective solution. Nevertheless, current P2P searching and routing arithmetic do not meet the requirement for extremely low time delay real-time multimedia application demand [7] so other network and parameters are to be examined. Figure 1 presents an example of an unstructured peer-to-peer overlay in comparison with the overall nodes participating in the physical network.

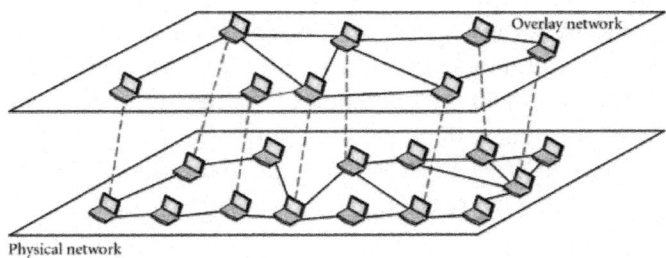

Figure 1: Diagram of an unstructured peer-to-peer overlay, green nodes participating in the overlay network.

In this paper we describe the design and prototype implementation of a communication platform aiming to provide voice and video communication in a distributed networking environment. Performance considerations and network characteristics have been taken into account in order to provide the set of properties dictated by the emergency and sensitive nature of the targeted application scenarios. The proposed system has been evaluated both by experimental means as well as subjective tests means. The results show that the proposed platform operates seamlessly in two hops, while in the four hops scenario, audio and video are delivered with marginal distortion. The conducted survey indicates that the user experience in terms of Quality of Service obtained higher scores in the scenario with the two hops.

The rest of the paper is organized as follows. In Section 2 we present the basic system architecture, while in Section 3 we present the results regarding the objective performance evaluation and the subjective survey-based user rating of the platform. Finally, Section 4 concludes our paper.

SYSTEM ARCHITECTURE

Network Organisation

A primary decision that needs to be made in defining the proposed communication platform is the type of underlying network organization. The autonomic nature of the network, where each node operates in a standalone fashion, is matched by several network organization paradigms, namely, peer-to-peer networks, mobile ad hoc networks, and so forth.

Given that the application focus of the proposed communication platform is on real-time multimedia provisioning, a set of basic properties of the underlying network organization needs to be met. To begin with, ease of deployment is necessary so that the network is quickly set up in a straightforward manner. Depending on the conditions on the deployment area, the network topology should be formed using simple procedures without requiring much computational or communicational effort. Furthermore, the decisions made about the construction and the operation of the network should be distributed across the network nodes, thus avoiding single (or a limited set of) point of failure. Resiliency in order to overcome potential failures of nodes is another requisite of the underlying network.

By definition, mobile ad hoc networks (MANETs) satisfy all the aforementioned properties. MANETs consist of wireless network devices that operate without any kind of centralized control or fixed communication infrastructure. Each network node operates not only as a host but also as a router forwarding packets to other nodes, which may not be within direct transmission range to each other. Therefore, each packet is transmitted to its destination in a multihop manner. The autonomous nature of MANETs fits the required properties of the real-time multimedia provisioning, while in the same time posing interesting challenges in defining efficient protocols in this direction.

Peer-to-Peer MANETs

In order to efficiently provide real-time multimedia transmission over MANETs, some notions of the formation and operation of peer-to-peer (p2p) networks are employed. Such networks are constructed based on an overlay network topology that is formed on top of the actual (physical) network topology and dictates the way the network peers are logically connected between each other.

Generally, three classes of peer-to-peer overlay networks related to the structure of the topology create structured, unstructured, and hybrid peer-to-peer architectures. In structured overlays, the architecture is controlled in an organized manner with content distributed at specific locations across the network to increase the efficiency of lookup queries. In unstructured overlays, the network is organized randomly in either a flat or hierarchical manner and execute queries using flooding, random walks, or expanding-ring Time to Live (TTL) techniques. Each peer receiving the query will initiate the search on its own local content which allows the execution of more complex queries. Finally, in hybrid peer-to-peer architectures, queries are handled by a central server which contains a database of content and its location within the overlay network, peers lookup the content on the main server then connect to the peer containing the specific content using the overlay. Peers are responsible for contacting the main server with the information on which resources and content they wish to share, making them an unfavorable solution on MANETs due to the single point of failure.

In order to setup, maintain and tear down multimedia sessions between network nodes, clearly define procedures for establishing, using, and terminating a logical connection between the terminal nodes. These procedures are part of a signaling protocol that in the proposed communication platform is based on the P2PSIP protocol [8]. P2PSIP is a peer-to-peer approach of the Session Initiation Protocol (SIP) [9] communication protocol that enables solutions for distributed storage of user information such as registration info along with logical position within the overlay and then handles all possible user queries leading to real-time communication session establishment. The most essential component of P2PSIP architecture is no other than the Distributed Hash Table (DHT). Its distributed nature originates on the fact that it is divided into several parts each located inside an overlay peer and stores the physical addresses of all participating nodes in order to use them for resource availability lookup. Each node that receives a query for the address of a certain node searches its DHT fraction and if it contains the requested info, it returns it to the node that posted the query, otherwise it forwards the request to its logical neighbors [10].

Routing Protocols

Due to the multihop nature of the packet forwarding in MANETs, routing is a primordial task directly affecting their performance. Therefore, efficient routing mechanisms are integral to a communication platform providing real-time multimedia in this context.

Designing a routing protocol for MANETs has been a very active research field in the recent years. The challenges that must be met are both numerous and diverse. The frequently changing topology is among the basic factors that must be taken into account, since the routes calculated at every time instance are subject to repeated change. Another priority is energy conservation, since, in the general case, the lack of fixed infrastructure means that energy recharging in network nodes is very difficult or costly, if not impossible. Furthermore, attention should be paid so as to evenly distribute the traffic across the network.

Traditionally, the basic categorization of routing protocols in MANETs is made based on whether their operation is proactive or reactive (on-demand). A protocol is defined as proactive when the routes from every node to everyone else are calculated and updated in a periodic fashion, while in the on-demand case a route is obtained only upon request from a packet. Both categories have advantages and disadvantages. Depending on the special characteristics of the network deployment setting, each of the aforementioned types of routing protocols may be suitable. Typical representatives of the proactive protocols are DSDV [11] and OLSR [12], while DSR [13] and AODV [14] are classic examples of on-demand protocols.

In the proposed communication platform a hybrid routing protocol is employed, ChaMeLeon (CML) [15]. The main concept of CML is the adaptability of its routing mechanisms according to the changes in the network topology. More specifically, it consists of 3 phases of operation, namely proactive, oscillation, and reactive. The basic criterion for the operation type selection is the network size. In relatively small networks, routing is implemented in a proactive fashion using the OLSR protocol, whereas when the number of nodes grows larger, CML utilizes the reactive Ad hoc On-Demand Distance Vector (AODV) [14].

Architecture Design

While attempting to implement a certain prototype that takes under consideration the latest trends in all aforementioned signaling, architecture, and routing protocols there were certain issues to be addressed mostly regarding the actual communication scheme amongst all participating entities. For being totally clear prior to the implementation the signaling diagram presented in Figure 2 was introduced.

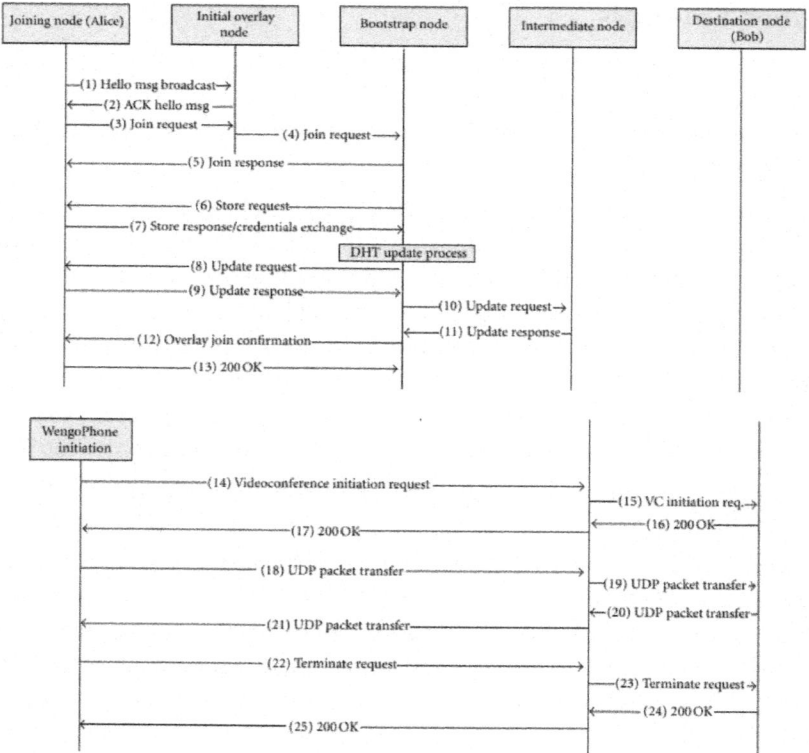

Figure 2: Join and single-hop communication signaling diagram.

There are five individual entities in this signaling diagram, the Joining Node (JN) which tries to access the overlay for the first time, the Initial Overlay Node (ION) which acts as the point of entry of the JN, the Bootstrap Node (BN) which operates as the key entity when it comes to DHT and overlay access and update, the Intermediate Node (IN) which plays the role of the packet carrier in the communication process and last but not least the Destination Node (DN) with whom the JN tries to establish communication in the first place. Normally, between the JN and DN there are several peers acting as IN thus facilitating communication. In all the tested scenarios described in the following paragraphs, more than one intermediate node is present. Nevertheless, for simplicity reasons for this particular diagram only one hop in the overall communication process between JN and DN is illustrated.

The joining peer first initiates a neighbor discovery mechanism in order to detect if there is someone within its communication range. JP broadcasts a HELLO message that is identified by every peer already connected to the overlay and receives an ACK HELLO message as response. Then JP transmits a JOIN Request to the ACK HELLO message sender, which in this particular case is the ION. After receiving the JOIN Request, ION forwards it towards the Bootstrap Node. BN then issues a JOIN Response towards the JN along with a STORE Request containing the public keys of JNs' logical neighbors, while JN confirms the successful reception of the later with a STORE Response message. Bootstrap then issues UPDATE Requests towards both JN and its logical neighbors (in the particular case the IN) informing them that they are connected to each other, in a message that contains the local view of the JP for the overlay. Finally, BN asks for overlay join confirmation by the JN receiving a 200 OK message in response.

Communication between JN and DN will include video and sound over IP network. Therefore after the software described in Section 2.5 becomes operational, JN sends a request regarding Videoconference (VC) initiation. The particular request is being forwarded through IN to the DN in a single-hop path which will be used for all signal and packet transfer for the whole session. DN responses also travel through the particular single-hop path. After JN receives the 200 OK message to its VC request the actual call begins. Through the monitoring software, the whole call was identified as UDP packet traffic, a common method used in all VoIP and real-time communication schemes. After the conclusion of the call JN sends a Termination request and the overall process finishes through a 200 OK message from the DN.

Extended WengoPhone Prototype Implementation

A prototype implementation was a matter of utmost importance for the necessary proof of concept of the solution we propose in this paper regarding real-time multimedia communication over ad hoc networks. With voice and video being the two most fundamental elements of human interaction, the prototype was designed in such a way that endorses both attributes in a seamless binding. Voice is integrated by the VoIP capability, the platform included and the additional feature of videoconference support were added to further improve Quality of Service and user experience. The implemented software utilized libraries and repositories of a certain project called WengoPhone [16], a SIP compliant VoIP client developed under the GNU General Public Licence (GPL). In addition, we extended the provided features of the software by developing tools that initiate a mobile device's web camera, capture video, and encode it using the H.263 video compression algorithm [17]. This algorithm is considered to be optimized for video transmission over wireless networks and also published under open source licence, unlike it's successors that have many patent limitations blocking us from exploiting all their attributes. The latest version of WengoPhone with the new enabled features has a redesigned Graphical User Interface (GUI) that is shown in Figure 3.

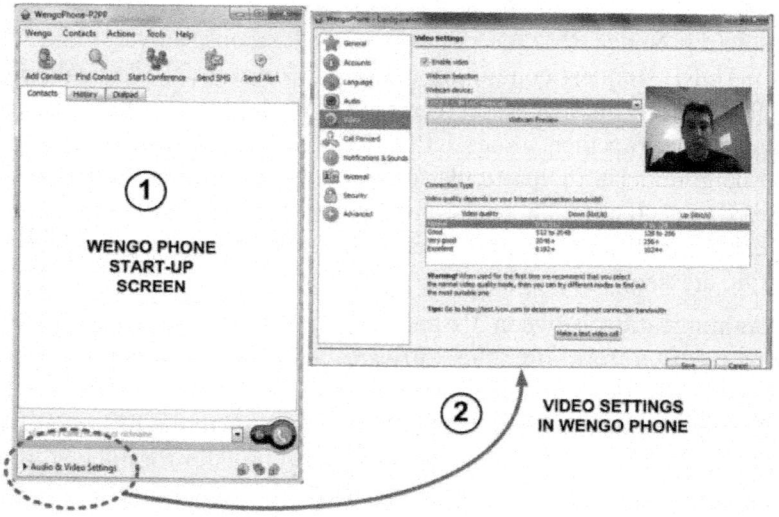

Figure 3: WengoPhone startup screen and video settings.

Providing that video quality is a key feature, we decided to include several quality levels that can be accessed instantly by a menu located in the bottom of the graphical interface presented in the previous figure. The new window for video settings is shown in the previous picture. After setting all attributes in order according user's preferences along with wireless network's capabilities, a video call is performed as shown in Figure 4.

Figure 4: Video call on WengoPhone software.

The application was developed using Microsoft Visual Studio 2005 and was tested on three different Asus EEEPCs having the followin specifications: Atom CPU running at 1.6 Ghz, 1 Gb RAM, 160 Gb HDD, and Windows 7 Professional Edition, each representing a mobile device of an extreme emergency communication scenario. For the GUI, Qt, a cross-platform application framework was used by installing a certain widget toolkit, since several alterations were considered necessary in terms of user friendliness, for providing direct access to the new features. No external video camera or wireless card was used apart those provided by the EEEPCs, a 1.3 Megapixel Logitech Camera and an Atheros 802.11 g/n compatible interface. A major concern during Wengo Phone extensions development was how to abandon the monolithic architecture the previous version had and move towards a more flexible platform that would make future changes and code upgrades easier. We manage to achieve this goal by keeping the H.263 algorithm implementation relatively modular, avoid to temper with preexisting pieces of code when possible, apart in case that this was absolutely necessary.

RESULTS

The software prototype described in a previous paragraph requires a whole testbed for extended evaluation of the new abilities it supports. In particular six wireless nodes forming an ad hoc network were used in order to check voice and video quality. In all our tests, although we expected that this relatively demanding platform would consume all available CPU and memory capacity of a low-range piece of equipment such as an EEEPC, CPU utilization never exceeded 40% and memory was constantly under 15% of its total capacity. The topology used for our tests is illustrated in Figure 5.

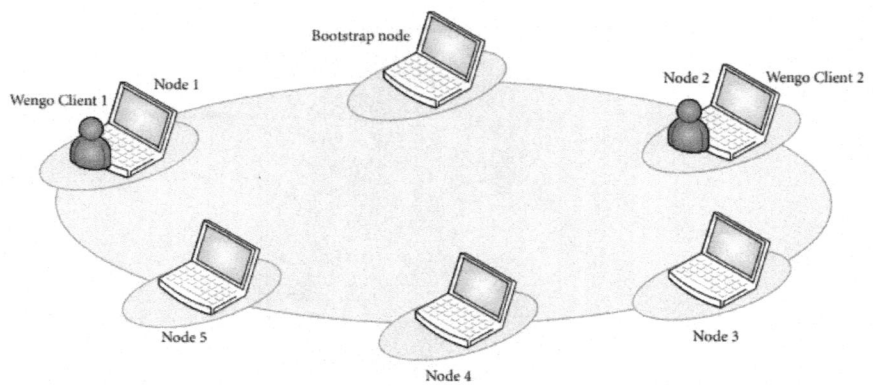

Figure 5: Testbed topology.

Only two out of six nodes had Wengo Phone software installed and they are depicted clearly in all pictures as WengoClient 1 and 2. Most nodes forming the necessary ad Hoc network use Windows 7 as their operating system, together with a compatible OLSR implementation for routing purposes. There is one exception the node that plays the role of the Bootstrap Node which is running Ubuntu Linux. Static IP addresses were given to all nodes in order to ensure that no interference from any foreign network will compromise the evaluation. OLSR and CML protocols were configured in such a way that a certain amount of hops between nodes to be established, according the two evaluation scenarios presented in the following section.

Evaluation Scenarios

For evaluating the performance of the implemented software, two scenarios were designed. Nodes 1 and 2 having Wengo Phone installed are trying to establish a call to each other. In order to achieve that call, they first have to join the overlay which is maintained by the Bootstrap Node. The main difference in these two scenarios is the actual routing configuration in OLSR and CML protocols. In the first scenario all IP packets involved in the call between Node 1 and Node 2 are being diverted through Node 3, achieving a total amount of two hops whithin the network overlay until they reach the destination node, Node 2 and 1, respectively. This is expected to keep the packet loss and the jitter relatively low thus providing a service as close to direct connection calls as possible. In the second scenario the OLSR and CML configuration has changed dramatically. Instead of a total of two hops, IP packets from Node 1 to Node 2 are being routed through all overlay nodes which means a four-hop route. This extensive rerouting is likely to increase packet loss and jitter causing issues in video quality due to image distortion. In order to have a more complete estimation regarding the performance of our software implementation in terms of video delivery under all conditions, two sets of measurements per scenario were taken. During the first one, video quality was set to "Normal" requiring available bandwidth of 0–512 Kbit/sec for upload and 0–128 Kbit/s for download, while for the second the video quality was increased to "Very Good" requiring available bandwidth of 512–2048 Kbit/sec for upload and 128–256 Kbit/s for download, quite an increase compared to the previous scenario. For monitoring all traffic between nodes we used the Wireshark network analyser [18]. In both scenarios the actual length of each call was set to 120 sec, enough time frame for measuring all network characteristics.

Experimental Results

Real-time applications such as VoIP and Video conferencing platforms are extremely sensitive in terms of jitter and packet loss. Packet loss compromises voice and video quality since data flow from the source to the destination is interrupted instead of being a continuous event, time slot is expired and the communication becomes unbearable. System cannot have the luxury to wait for a retry as in other applications and users do not experience best possible Quality of Service. Jitter refers to undesired variation in packet receiving. If there is a traffic delay, data might be buffered accordingly but when this delay keeps accumulated, buffer can no longer sustain packet delivery and this could result in video distortion or jerkiness. When it comes to voice jitter it causes gaps in communication and problematic system behaviour in general.

We used Wireshark [18] network analyser in order to monitor and record traffic and packet exchange between nodes as shown in Figure 6. In both scenarios, Wireshark daemon was strategically placed so that traversal packet flow can be logged, thus creating a file for further analysis. Focusing on Jitter and Packet Loss we were able to compare the overall testbed behaviour, not only the improved version of WengoPhone we developed. Our analysis brought to light several interesting results regarding multihop topologies and how the hop number increase inevitably leads to voice and video quality deterioration.

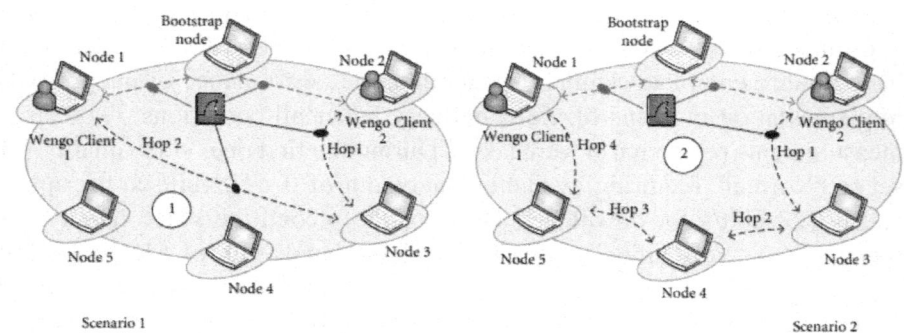

Figure 6: Evaluation scenarios.

Figure 7 presents jitter measured when a video-call session was established between Nodes 1 and 2 in each of the scenarios presented earlier. Video quality is set to "Normal" in both sessions making number of hops their only difference. In Scenario 1, when only two hops are needed for a UDP packet to cross the gap between source and destination, jitter never exceeded 60 ms in absolute numbers. On the other hand, in Scenario 2, with four hops between source and destination, jitter had an average value of 81 ms, significantly increased compared to the previous topology, yet acceptable.

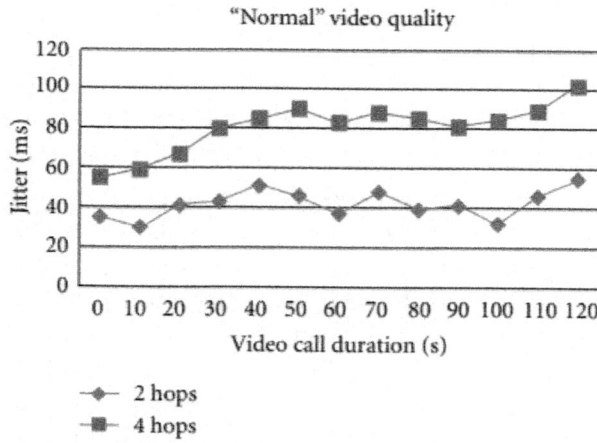

Figure 7: Jitter when video quality is set to "Normal".

We measured jitter again, this time after terminating all previous calls and restarting WengoPhone setting video quality of the established video-call session between Nodes 1 and 2 to "Very Good". This time jitter values were generally increased due to higher traffic caused by packet increase in each topology. The results are shown in Figure 8, with jitter in Scenario 1 having a mean value of 64.3 ms while that of Scenario 2, with four hops end-to-end, rises well above 110 ms. A jitter value greater than 100 ms urges for buffer implementation, a temporary fix sometimes of no use in real-time applications.

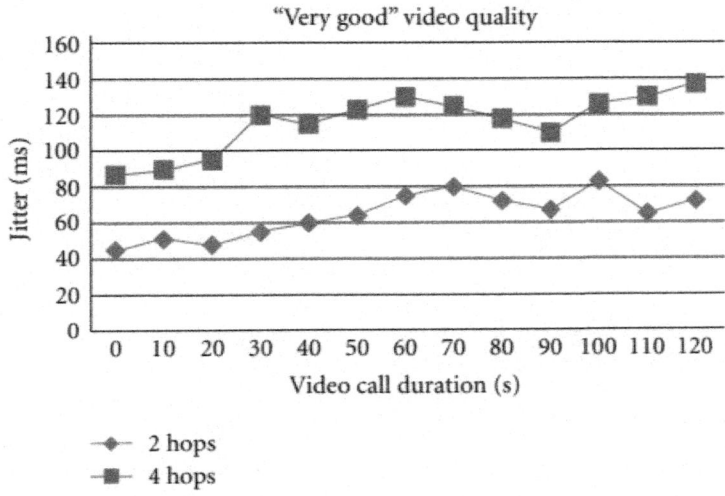

Figure 8: Jitter when video quality is set to "Very Good".

In addition to jitter, Packet Loss was measured for all previously mentioned scenarios and video quality settings. The results are shown in Figures 9 and 10.

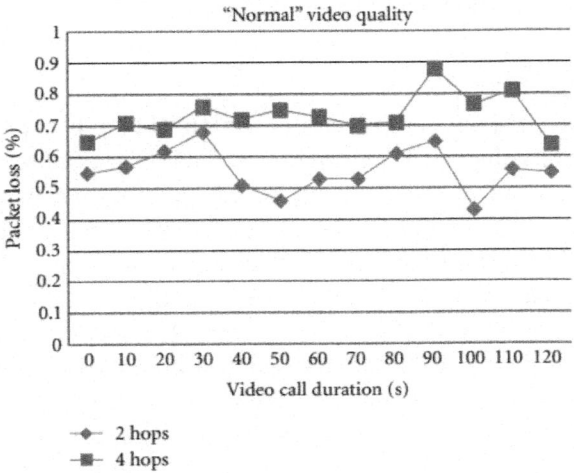

Figure 9: Packet loss when video quality is set to "Normal".

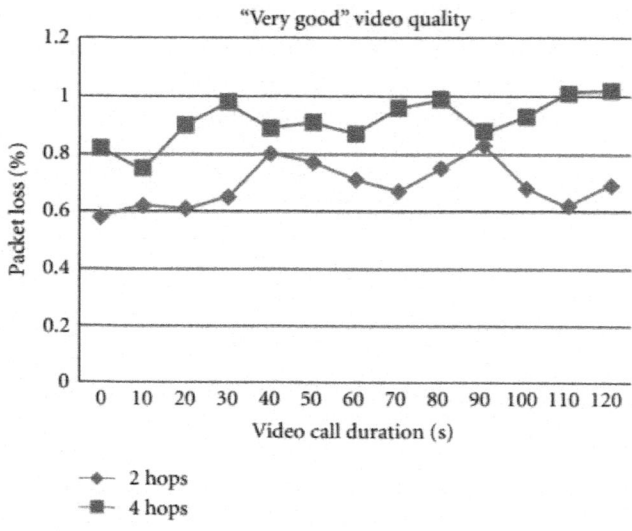

Figure 10: Packet loss when video quality is set to "Very Good".

Typical numbers for acceptable packet loss during a videoconference system range from 0.1% to 1%. Video is less sensitive than voice. According to VoIP standards packet loss greater than 1% is likely to compromise the whole session, leading to several audio drop-outs [2]. This inevitably draws the conclusion that in the four-hop routing of Scenario 2 users might experience problems, since packet loss percentage rises to 1.03% after less than 120 sec of established audiovisual communication.

Subjective Evaluation

The International Telecommunication Union (ITU) has published several recommendations that intend to define standards for subjective assessment methods to be used in the one-way overall audiovisual quality evaluation. These methods can be used for several different purposes, including but not limited to ranking of audiovisual system performance and evaluation of the quality level during an audiovisual connection. The most commonly used ITU Recommendation is P.911 [19], where several test methods and experimental design techniques are presented. Although a number of such methods have been validated for different purposes, the final choice of one of these methods for a particular application depends on various factors such as the context, the purpose, and where in the development process the test is to be performed. Out of all proposed methods described in [19] the most suitable for our evaluation seems to be the Absolute Category Rating (ACR). More information regarding this method as well as the numerical evaluating scale we used can be found in Appendix.

Subjects

The possible number of subjects in a viewing and listening test along with usability tests on terminals or services varies from 6 to 40. Four is the absolute minimum for statistical reasons, while there is rarely any point in going beyond 40 [19]. The actual number in a specific test should really depend on the required validity and the need to generalize from a sample to a larger population. In general, at least 15 subjects should participate in the experiment. They should not be directly involved either in picture or audio quality evaluation as part of their work and should not be experienced assessors. Prior to the session the observers should usually be screened for normal color vision as well as normal or corrected-to-normal acuity. During our evaluation 32 individuals were divided in two groups. Each group participated in videoconference sessions in random pairs without knowing the total amount of intermediate hops packets were routed into, for not making them provide relatively biased overall experience evaluation.

After a videoconference session of 120 sec was concluded, subjects were given a form containing a set of four standardized questions for the aforementioned rating. These questions were as follows.(1)How would you rate the video quality of the connection?(2)How would you rate the audio quality of the connection?(3)How would you judge the effort needed to interrupt the other party?(4)How would you rate the overall audiovisual quality?

Answers for questions one, two, and four were given on the nine-level scale with greater numbers indicating more assigned points ergo proportional to user satisfaction. The third question was evaluated in a different way; since in a perfect video-call little effort would be sufficient for one participant to interrupting the other, points were assigned disproportional. More points indicate less effort therefore better user experience.

Subjective Evaluation Results

In Figure 11 the results of video quality evaluation are shown. A significant fact is that although video quality was set to "Normal" in both cases, there is little yet existing variation in user scores. This has a very profound explanation. Video quality in our testbed depends not only on the software prototype configuration but to the architecture and topology parameters as well. This means that video sessions of better quality that might reach to their destination distorted, delayed, or jerked due to jitter or latency is likely to satisfy users less than lower quality ones, scoring lower than expected. In our evaluation this seemed to be the case.

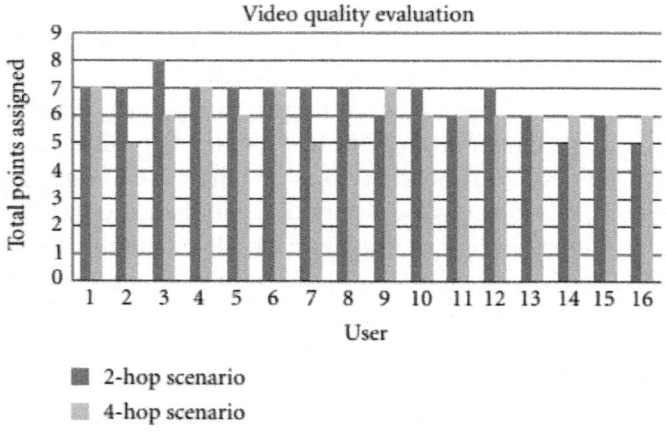

Figure 11: Video quality evaluation.

Audio quality evaluation results are shown in Figure 12. Once again network parameters seem to play a vital role to user satisfaction, since in almost all cases scenario with two hops leading to lower jitter and packet loss prevailed in user preferences.

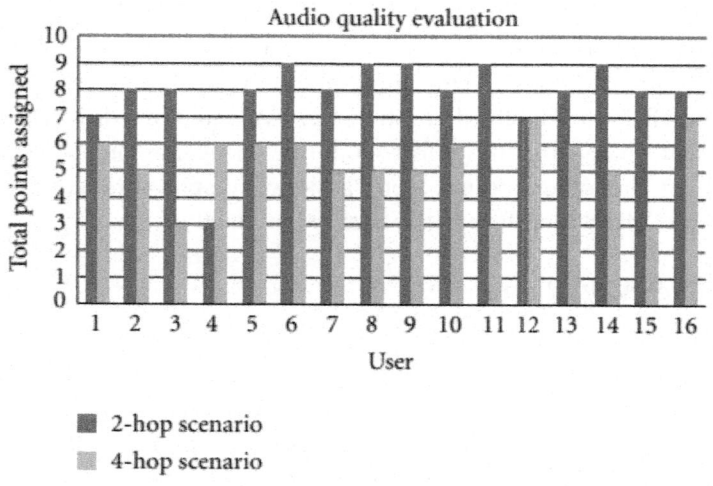

Figure 12: Audio quality evaluation.

In Figure 13 interruption effort results are depicted. Once again there is a slight user preference for the two-hop topology. This might be explained if we consider the fact that although video is involved in the conference, the most natural way of human communication is verbal and when this is compromised for instance due to jitter, subjects feel uncomfortable and assign points accordingly.

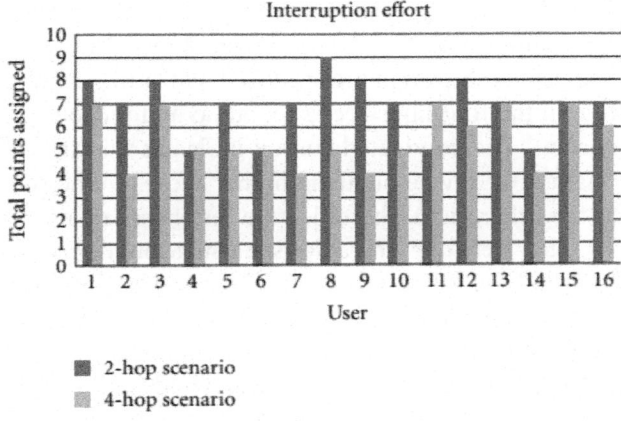

Figure 13: Interruption effort needed during video conference.

In Figure 14 overall audiovisual quality evaluation can be found. Scenario 1 once again overcame Scenario 2 in terms of user satisfaction proving that in video-conference applications as well as in all real-time communication platforms, network topologies having better jitter and packet loss ratings shall gain momentum over more complex but less effective ones.

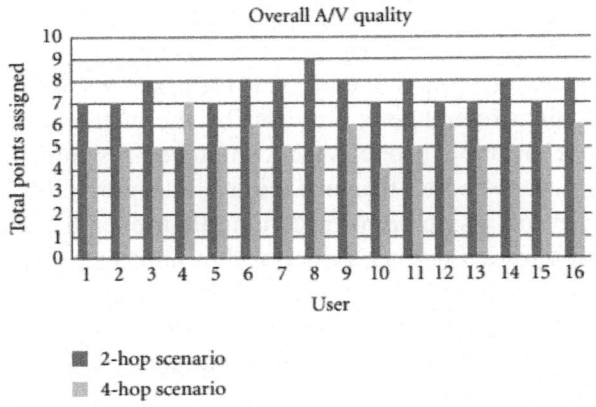

Figure 14: Overall audio/video quality evaluation.

CONCLUSIONS

The task of evaluating such a complex ad hoc network platform including a real-time communication prototype proved to be extremely challenging. Using all necessary software tools as well as modern industry standards we were able to perform both subjective and objective evaluation of our proposed solution. Two different scenarios were designed, based on network characteristics as well as prototype restrictions. In the first one packet routing was delivered in a total of two hops end-to-end, while in the second one a total of four hops was introduced. The results show that the proposed platform operates seamlessly in two hops, while in the four hops scenario, audio and video are delivered with marginal distortion. The conducted survey indicates that the user experience in terms of Quality of Service obtained higher scores in the scenario with the two hops. On the other hand, the objective test bears many consistencies to the subjective one. Network characteristics measurements acquired during the test indicate that the key elements of jitter and packet loss were slightly compromised in the four-hop scenario, a fact that shows clearly the limitations this topology has until today. Future research is definitely going to improve all characteristics of the mobile ad hoc networking and we hope this paper, based on an implementation rather than simulation results, paved a path towards that direction.

APPENDIX

A. Absolute Category Rating

The Absolute Category Rating method is a category judgment where the test sequences are presented on at the time and are rated independently on a category scale. In our case, instead of a video/audio sequence, the evaluation refers to the videoconference and its parameters. The method specifies that after each videoconference session, the subjects are asked to evaluate the session's quality. No explicit reference is provided by the method, although subjects will always use an implicit one. Videoconference session can be no longer than 2 minutes yet no shorter than one minute, thus providing enough time for user to consider and assign points according their overall experience. This is persistent with our measurment scenarios that limit the total call duration to 120 sec. ITU recommends a five-level scale for rating overall quality [19]. In our evaluation since higher discriminative power is required because of the low bit rate of the video conference encoding algorithm, the nine-level scale presented in Table 1 is going to be used.

Table 1

9-8	Excellent
7-6	Good
5-4	Fair
3-2	Poor
1	Bad

Additional examples of suitable numerical or continuous scales are given in [19], which also provides examples of rating dimensions other than overall quality. Such dimension may be useful for obtaining more information on different perceptual quality factors when the overall quality rating is nearly equal for certain systems under test, although the systems are clearly perceived as different. In our case the simple nine-level scale is considered rather accurate. For the ACR method, the necessary number of replications is obtained by repeating the same test conditions at different points of time in the test. ACR is easy and fast to implement and the presentation of the stimuli is similar to that of the common use of the systems. This attribute renders ACR an optimal choice for qualification tests.

REFERENCES

1. M. C. Castro, A. J. Kassler, C. F. Chiasserini, C. Casetti, and I. Korpeoglou, "Peer-to-peer overlay in mobile Ad-hoc networks," in Handbook of Peer-to-Peer Networking, Springer, 2010.

2. A. Mawji, H. Hassanein, and X. Zhang, "Peer-to-peer overlay topology control for mobile ad hoc networks," Pervasive and Mobile Computing, vol. 7, no. 4, pp. 467–478, 2011.

3. B. G. Chun, R. Fonseca, I. Stoica, and J. Kubiatowicz, "Characterizing selfishly constructed overlay routing networks," in Proceedings of the 23rd Annual Joint Conference of the IEEE Computer and Communications Societies, pp. 1329–1339, Hong-Kong, March 2004.

4. C. Jennings and D. A. Bryan, "P2P for communications: beyond file sharing," Business Communications Review, vol. 36, no. 2, p. 36, 2006. ·

5. M. Cavalcanti de Castro, Enabling multimedia services over wireless multi-hop networks [M.S. thesis], Department of Computer Science, Karlstad University, 2009.

6. Y. Kim, Y. Kim, and N. Kang, "Multimedia push-to-talk service in home networks," IEEE Transactions on Consumer Electronics, vol. 56, no. 3, pp. 1480–1486, 2010.

7. Y. Mo, F. Wang, B. Huang, and F. Luo, "Real-time multimedia services using hybrid P2PSIP," inProceedings of the International Symposium on

Communications and Information Technologies (ISCIT ‹06), Bangkok, Thailand, 2006.

8.   P2PSIP, http://www.p2psip.org/ietf.php.

9.   H. Schulzrinne, G. Camarillo, A. Johnston et al., "SIP: session initiation protocol," RFC 3261, 2002.

10. C. Tselios, K. Birkos, P. Galiotos, T. Dagiuklas, and S. Kotsopoulos, "Malicious threats and novel security extensions in P2PSIP," in Proceedings of the 2nd International Workshop on Pervasive Networks for Emergency Management, Lugano, Switzerland, March 2012.

11. C. E. Perkins and P. Bhagwat, "Highly dynamic destination-sequenced distance-vector routing (dsdv) for mobile computers," in Proceedings of the Conference on Communications Architectures, Protocols and Applications, pp. 234–244, 1994.

12. T. Clausen and P. Jacquet, "Optimized Link State Routing Protocol (OLSR)," IETF RFC, 3626, Network working Group: Project Hipercom, INRIA, October 2003.

13. The Dynamic Source Routing Protocol for Mobile Ad Hoc Networks (DSR), 2003,http://tools.ietf.org/html/draft-ietf-manet-dsr-10.

14. C. Perkins, E. Belding-Royer, and S. Das, "Ad hoc on-demand distance vector (AODV) routing," RFC3561, 2003.

15. T. Ramrekha, E. Panaousis, and C. Politis, "ChaMeLeon (CML): A hybrid and adaptive routing protocol for Emergency Situations," IETF internet draft, http://tools.ietf.org/html/draft-ramrekha-manet-cml-01.txt.

16. The WengoPhone VoIP client, http://wengophone.en.softonic.com/.

17. ITU-T, "H.263: Video coding for low bit rate communication," 2010, http://www.itu.int/rec/T-REC-H.263/.

18. The Wireshark Packet Analyzer, http://www.wireshark.org/.

19. ITU-T, "Recommendation P.911, Subjective audiovisual quality assessment methods for multimedia applications," 1998.

Chapter 3

INTEGRATING USAGE CONTROL WITH SIP-BASED COMMUNICATIONS

E. Barka[1] and A. Lakas[2]

[1]Information Security Track, College of Information Technology, United Arab Emirates (UAE) University, Al Ain, United Arab Emirates

[2]Networking Track, College of Information Technology, United Arab Emirates (UAE) University, Al Ain, United Arab Emirates

ABSTRACT

The Session Initiation Protocol (SIP) is a signaling protocol used for establishing and maintaining communication sessions involving two or more participants. SIP was initially designed for voice over IP and multimedia conferencing, and then was extended to support other services such as instant messaging and presence management. Today, SIP is also adopted to be used with 3G wireless networks, thus it becomes an integral protocol for ubiquitous environment. SIP has various methods that support a variety of applications such as subscribing to a service, notification of an event, status update, and location and presence services. However, when it comes to security, the use of wireless and mobile communication technologies and the pervasive nature of this environment introduce higher risks to security than that of the old simple environment. In this paper, we introduce new architecture that implements a new type of access control called usage access control (UCON) to control the access to the SIP-based communication at preconnection, during connection, and postconnection. This will enable prescribers of SIP services to control who can identify their locations to approve or disapprove their subsequent connections, and to also set some parameters to determine whether a certain communication can continue or should terminate.

INTRODUCTION

Nowadays, our society is impacted by a revolutionary innovations in information technology that made communication around the globe seems like it is only a mile away. With these advances in technology, particulary in communications, we are also encountering a series of new problems on security and privacy issues.

Among these technologies are the protocols known as signaling protocols, which are used to carry call setup information. These protocols are used to set up, control, and maintain sessions for multimedia applications. Applications which require sessions to be set up include telephony, videoconferencing, and remote learning. SIP signaling protocols are used to set up sessions over packet-switched networks for IP-based multimedia applications such as voice-over IP (VOIP) and IP-based videoconferencing.

SIP has several methods that support a variety of applications such as subscribing to a service, notification of an event, status update, and location and presence services.

SIP provides flexible and real-time communications in a ubiquitous way which adds additional risks that result from the adoption of ubiquity, mobility, and heterogeneity.

Traditional access controls typically focus on the protection of data in closed environments, and the enforcement of control has been primarily based on identity and attributes of a known user. These types of access control lack a comprehensive, systematic approach to fulfill the security requirements of today's pervasive and ubiquitous nature.

To address these issues, we introduce in this paper a new architecture that implements a new type of access control called usage access control (UCON) to control the access to the SIP-based communication at preconnection, during connection, and postconnection. This will enable prescribers of SIP services to control who can identify their locations to approve or disapprove their subsequent connections, and to also set some parameters to determine whether a certain communication can continue or should terminate.

The remainder of this paper is organized as follows. In Section 2, we present the main motivation in using the model of UCON for SIP-based communications. Section 3 provides some background on SIP and discusses its role in the pervasive environment from a communication perspective. We do this by first providing an overview of SIP applications and related security aspects. In Section 4, we introduce the next generation access control, depicted in the usage control model (UCON). Section 5 presents our architecture and explains the integration of SIP with UCON. Finally, Section 6 discusses future work and concludes our paper.

MOTIVATION

In this section, we present few factors that support the usage control over the traditional access control through the use of UCON model in the case of SIP applications. Traditional access control models that are currently used for SIP communications suffer from the lack of flexibility to specify security policies in the case of ubiquitous environments.

For example, in the case of pay-per-usage, there is a need to terminate the SIP communications when the caller is out of credit. Traditional access control provides mechanisms only to control access during the establishment of an SIP call session, but completely loses control over the session during the call (when user credentials may change), and after the session has terminated. In this case, no control is provided after the access is granted even when the caller runs out of credit. This problem has been partially solved by introducing the back-to-back UA (B2BUA) mechanism to SIP architecture by forcing all communications to go through the SIP server including the media connection. However, this approach created other issues including the creation of a bottleneck situation, and generated additional processing overhead at the server. UCON can solve this issue, by specifying policies that monitor the SIP communications before and during the call. Moreover, it mandates and enforces continuous compliance to access conditions. In the case of noncompliance to these conditions, UCON provides mechanisms for revoking the access. Monitoring and enforcement mechanisms are now collocated and distributed within the SIP UAs.

In overall, UCON benefits consist specifically in its ability to specify and enforce usage policies not only before access, but also during the access and after.

SIP OVERVIEW

SIP, which is one of the most popular IP signaling protocols, is an application layer protocol created by the Internet Engineering Task Force (IETF) [1] to allow entities to locate one another on a network and invite themselves to participate in a session. It is also responsible for maintaining and terminating a session. SIP is limited only to the session establishments and terminations. Once the participating parties negotiate the characteristics of their communication, the session is established. The participating parties use the Session Description Protocols (SDPs) to define the audio or the video bearer channel they would like to communicate through. SDP is embedded within the SIP messages.

The architecture of SIP is based on a set of components which content varies based on the application deployed. The basic set of SIP components includes the following.

1.   User agent (UA): works on behalf of users to set up calls and establishes a multimedia communication.

2.   Proxy servers (PSs): keeps track of location of UAs and facilitates the establishment of sessions between them.

3.   Registrar: UAs keep the registrar updated on their current locations by initially registering to it with their current location. UAs also update the registrar with their preferred reachability information.

Figure 1 illustrates the basic components of SIP.

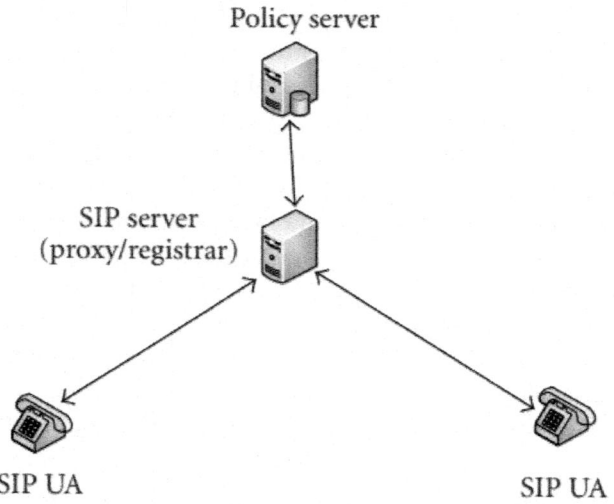

Policy server

SIP server
(proxy/registrar)

SIP UA   SIP UA

Figure 1: SIP communication.

SIP Applications

SIP provides support for various text, voice, and video session-based applications. The most popular application is VoIP. SIP comes natively with its own telephony applications; however, it may be used for the control and the management of IP telephony of various systems. For instance, SIP has been adopted by the 3GPP initiative to control and manage the communication between 3GPP components for the 3G and 4G telephony systems [2].

Another application for SIP is the instant messaging and presence management [3]. Instant messaging (IM) is a text-based communication that has seen an increasing popularity on the Internet. IM is accompanied with the presence of functionality where any IM communication is assorted with information which includes status and location information such as that contained in "buddy" list. This extension includes functionalities such as

1. publishing and uploading of presence information;
2. presence and event notification;
3. delivering of instant messages.

SIP Security

SIP provides flexible and real-time communications in a ubiquitous way. However, with doing so, it also adds extra risks due to the new factors of ubiquity, mobility, and heterogeneity. Some of the threats that are inherent to the use of SIP are listed as follows.

1. Registration hijacking: a registrar assesses the identity of a UA. The From header of an SIP request can be arbitrarily modified and hence open to malicious registration.
2. Impersonating a server: a UA contacts a proxy server to deliver requests. The server could be impersonated by an attacker. Mobility in SIP further complicates this scenario.
3. Tampering with message bodies.
4. Tearing down sessions—insert a BYE.
5. Denial-of-service attacks—denial-of-service attacks focus on rendering a particular network element unavailable, usually by directing an excessive amount of network traffic at its interfaces. In much architecture SIP proxy servers face the public Internet in order to accept requests from worldwide IP endpoints. SIP creates a number of potential opportunities for distributed denial-of-service attacks that must be recognized and addressed by the implementers and operators of SIP systems.

Therefore, the security challenges facing SIP are to ensure the following.

1. Authentication—SIP currently has the HTTP style digest mechanism, but it is not enough. A single sign-on authentication mechanism is needed.
2. Authorization using policy-based mechanisms—the read/write/ execute controls that are embedded in file systems. Some people have recommended, and tried to implement, traditional access control models, but they are broadly categorized as discretionary access control (DAC) [4, 5] and mandatory access control (MAC) models [4–6]. Others have proposed new models such as role-based access control (RBAC) and

task-based access control (TBAC) to address the security requirements [7].

The above mentioned solutions are not sufficient enough for providing security for pervasive environments that requires continuous control of the access "not just at the establishment of the connection" to the system resources by very heterogeneous types of users and devices; but also during and after a session is terminated.

Our approach introduces a new type of access control named usage access control (UCON) [8–10]. UCON is used here to control the access to the SIP-based communications: presession, during the session, and even at postsession. There are many benefits to UCON usage in SIP. This will enable subscribers to SIP services to control who can identify their locations, to approve or disapprove their subsequent connections, and to also set some parameters to determine whether a certain communication can continue or should terminate. UCON also contributes to solving some of the problems that are specific to some applications such as the necessity to have back-to-back UAs (B2BUA) [6]. The pay-per-usage application is an illustration of this problem. In this case, UCON may terminate the call when the caller runs out of credit.

The following section provides an overview of the usage control model (UCON).

USAGE CONTROL MODEL (UCON) OVERVIEW

In this section, we briefly review the general ideas of UCON and the core authorization models. The details of these models can be found in [8, 9].

The UCON model consists of six components, three of these componenets "subjects, objects, and rights" are considered core componenets and the other three "authorization rules, conditions, and obligations" are additional, and they are mainly involved in the authorization process (see Figure 2).

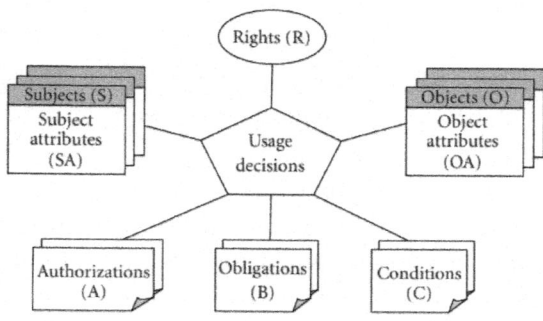

Figure 2: UCON components.

In UCON system, at least the authorization rules (specifically rights-related authorization rules) have to be included for authorization. Conditions and obligations can also be used in the authorization process. The following subsections describe UCON's core components.

Subjects

Subjects are "Active" entities associated with attributes, and hold and exercise certain rights on objects. Attributes are properties of the subjects that can be used for the authorization process. Examples of attributes include identities, roles, credits, memberships, security levels, and so forth. A subject can be a user, a group, a role, or a process. A user is an individual entity that has certain rights on an object. A group is a set of users who holds same rights as a group. A role is a named collection of users and relevant permissions [6]. Groups and roles may have hierarchical relationships.

Objects

Objects are "Passive" entities that subjects hold rights on, whereby the subjects can access or use objects. Objects are also associated with attributes, either by themselves or together with rights. As for subjects, the attributes include certain properties that can be used for the authorization process. Examples of object attributes are security levels, ownerships, classes, and so forth. Object classes are used to categorize objects so authorization can be done based not only on individual objects but also on sets of objects that belong to same class [8, 9]. In some cases, objects or objects with attributes (i.e., classes) are associated with attributes together with rights. Examples of the attributes for objects with rights are credits, roles, memberships, and so forth. The credits may be used to define how many credits are required to obtain a certain right on a specific object. For example, an e-book together with a read right may require $10 or the book with an additional print right may require $15.

Rights

Rights are privileges that a subject can hold on an object. Rights consist of a set of access/usage functions that enables a subject's access to objects. The authorizations of rights require associations with subjects and objects. Rights may or may not have a hierarchy.

UCON rights can be divided into many functional categories. The two most fundamental right categories might be a view and a modification. They are denoted as V and M, respectively, such that=>

$$R=\{V,M\}. \tag{1}$$

Modification includes change to an existing digital object and creation of a new object that reuses an original digital object. The range of V and M is denoted as

$$C=\{0,1,\alpha\} \tag{2}$$

where

1.   "0" means closed to everybody (no one can access);

2.   "1" means open to everybody (everyone can access); and

3.   "α" means access approval is selective or controlled.

The openness of the control or availability of object to public is expressed as $0<\alpha<1$ which means that 1 is most open to public and 0 is least open. The following subsections describe the additional component of UCON.

Authorization Rules

Authorization rules are a set of requirements that should be satisfied before allowing subjects' access to objects or use of objects.

There are mainly two kinds of authorization rules. The rights-related authorization rules (RARs), which are used to check if a subject has valid privilege to exercise certain rights on a digital object.

Examples of such rules may include identities or roles verification, properties checking, proof of payments, and so forth. Moreover, the obligation-related authorization rules (OARs), which are mainly used to check if a subject has agreed on the fulfillment of an obligation which has to be done after obtaining or exercising rights on a digital object. Examples of such rules may include metered payment agreement, usage log report agreement, and so forth.

The authorization rules are different from conditions in that the authorization rules are a set of decision factors used to check whether a subject is qualified for the use of certain rights on an object, whereas the conditions are decision factors used to check whether existing limitations and status of usage rights on an object are valid and whether those limitations have to be updated.

Conditions

Conditions are a set of decision factors that the system should verify at authorization process along with authorization rules before allowing usage of rights on a digital object.

Conditions are of two types: dynamic conditions are conditions which include information that may have to be checked for updates at each time of usage. Examples of dynamic conditions are the number of usage times (e.g., can read 5 times, can print 2 times), and usage log (e.g., already read portion cannot be accessed again). On the other hand, static conditions include information that does not have to be checked for updates. Some examples of static conditions are accessible time period (e.g., business hours), accessible location (e.g., workplace), and allowed printer name.

Obligations

Obligations are mandatory requirements that a subject has to perform after obtaining or exercising rights on an object. In real-world implementation, however, this may have to be done by agreeing on the fulfillment of obligations before obtaining the rights and at the time obligation-related authorization rules are checked. For example, a consumer subject may have to accept metered payment agreements before obtaining the rights for the usage of certain digital information or should agree on providing usage log information to a provider subject before reading an e-book or listening a music file. Traditional access control has hardly recognized the obligation concept.

The most important properties that distinguish UCON from traditional access control models and trust management are the continuity of usage decisions and the mutability of attributes. Continuity means that a control policy may be enforced not only before an access, but also during the period of the access.

The control decision components are checked and enforced in the first two phases, named, predecisions and ongoing decisions, respectively. In the after-usage phase, we do not enforce any policy since there is no access control after a subject finishes a usage on an object.

Mutability refers to updates of the subject or the object's attribute that may occur as a result of the access. Along with the three phases, there are three kinds of updates: preupdates, ongoing updates, and postupdates. All these updates are performed and monitored by the system.

An update of subject or object attributes may result in a system action to permit or revoke an access. An update can affect not only the concurrent usage, but also other usages related to the same subject or object. An update on the current usage may generate.

SIP/UCON INTEGRATION

Integrating the UCON technology into ubiquitous SIP-based environment requires a careful mapping between the entities of UCON and those entities and components of the SIP. Following is a list of integrated components which require such mapping:

User/Subjects

The concept of participants in SIP is represented as a user component in the UCON.

Permissions/Rights

The concept of permissions in UCON will reflect all the privileges that an SIP participant needs to complete a task.

Objects

The objects in UCON are used to represent all entities in SIP that an SIP UA's seek to connect to.

Authorization Rules

Authorization rules in UCON are the set of requirements that should be satisfied before any SIP UA and be permitted to establish any connection with any other SIIP entity.

Obligations

The concept of obligation in UCON can be represented in SIP as the set of actions that an SIP user is required to perform before and after the connection has been established.

Conditions

Conditions in UCON are represented in SIP by the set of environmental and system decision factors that must be continuously evaluated to make sure that their changes do not lead to changes in the connection status.

Architecture of the UCON/SIP

One of the most critical issues in using UCON for enforcing access into SIP environment is to use the concept of a reference monitor. The reference monitor (RM) has been introduced, and extensively discussed, by the access control community for years. The concept of a reference monitor was introduced and published by the ISO in a standard for access control framework [11]. The RM concept has been considered as the core control mechanism for access and usage of digital information. In classical access control, subjects can access digital objects only through the reference monitor, which is a process inside the trusted computer base that is always running and is a tamper proof.

The following section discusses our conceptual structure of UCON/SIP access control domains, based on the reference monitor.

UCON/SIP Areas of Control

The area of control in our architecture refers to the area of coverage where the rights to access the SIP objects is under the control of the reference monitor.

According to the standard [11], the reference monitor consists of two facilities: access control enforcement facility (AEF) and access decision facility (ADF). AEF and ADF interact such as every request to access an object is intercepted by AEF. The later asks the ADF for a decision on the request approval. ADF returns either "yes" or "no" as appropriate. The enforcement of this decision takes place at the AEF.

Our UCON/SIP reference monitor is similar but differs in the details from that of ISO. Figure 3 shows the conceptual structure of the UCON/SIP reference monitor.

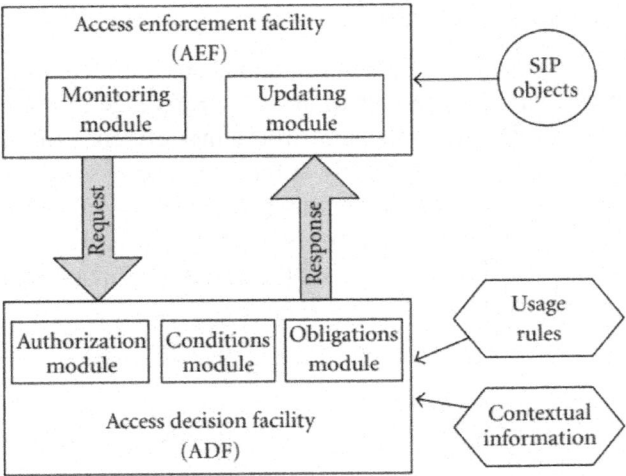

Figure 3: Conceptual structure for UCON/SIP reference monitor.

As the figure shows, both the AEF and ADF include several functional modules. AEF contains

1. monitoring module which is used to keep track of the changes of the attributes of the subjects and objects;

2. updating module which is used primarily to update the attributes of the subjects and objects.

The ADF is where all of the access granting decisions, the access maintenance, and the access termination take place. This facility includes three core modules that are utilized collectively in rendering access decisions as response to the (AEF) requests:

1.   authorization;
2.   condition;
3.   obligation.

The authorization module uses subjects and objects attributes, and the access rules to check if a request is allowed or not. The condition module uses the access rules and the contextual information to decide whether the conditional requirements, both system and environmental, for the authorized request are satisfied or not. Finally, the obligation module is responsible for handling decisions that are tied to actions that are required to be performed by the requestor before, during or after the access is granted. All existing obligations are monitored by the monitoring module and the outcome must be resolved by the update module in the AEF.

Areas of Control Architecture

To control the access to the SIP environment, our architecture considers two areas of controls, based on the location of the reference monitor. The server-side control domain (SCD) is the area where the reference monitor is located at the server environment and directly enforces the access policy to the system resources. The client-side control domain (CCD), on the other hand, is the area of control where the reference monitor is located at the client environment and enforces the access policy on behalf of the server. Figures 4 and 5 depict this architecture, respectively.

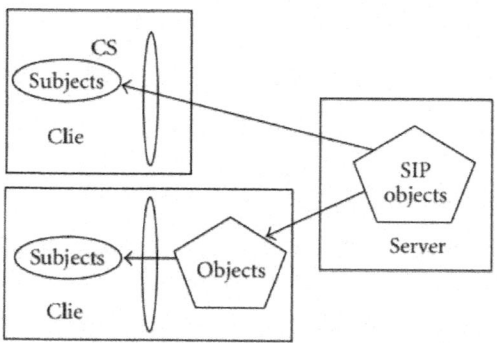

Figure 4: Client-side reference monitor.

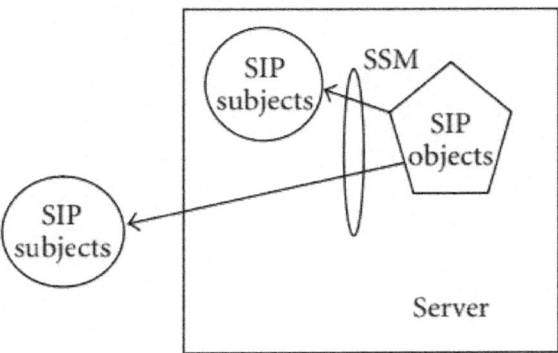

Figure 5: Server-side reference monitor.

Figure 4 shows that the reference monitor does not reside within the area of the server, but rather at the client side. This setup provides for better usage control over system objects. In this case, because of the existence of the SRM, the system objects can be stored centrally or locally, in either case, the objects are under the control of the client instead of the server.

Figure 5 shows that, in the case of the server-side control, the control of subject's access to objects is done centrally. Moreover, in this setup, the subject can either be located within the network or outside, and the SIP objects may or may not be stored in the client's storage, depending upon the criticality and sensitivity of the content of the object. If it is not that sensitive, then it can be allowed to reside outside of the server-side storage. However, if the content is very critical or very sensitive, the object must stay within the server-side storage.

Figure 6 illustrates the scenario of an SIP user agent client (UAC) initiating a call to a user agent server (UAS). This call is established through the SIP proxy. For this call to take place, the SIP proxy will verify the credentials of the caller and the callee. This scenario is executed through the following steps.

1. UAC initiates the call by sending an SIP INVITE to the SIP proxy.

2. The SIP proxy verifies the credentials of UAC and UAS by soliciting locally a decision from the ADF.

3. ADF looks up the access rules at the policy server, rules that are relevant to UAC and UAS, makes a decision on whether UAC is authorized to make call to UAS, and sends locally a notification to the AEF for the enforcement of such decision.

4. In the case of acceptance, the AEF notifies the proxy with the decision which will allow the proxy to proceed with the processing of the call establishment.

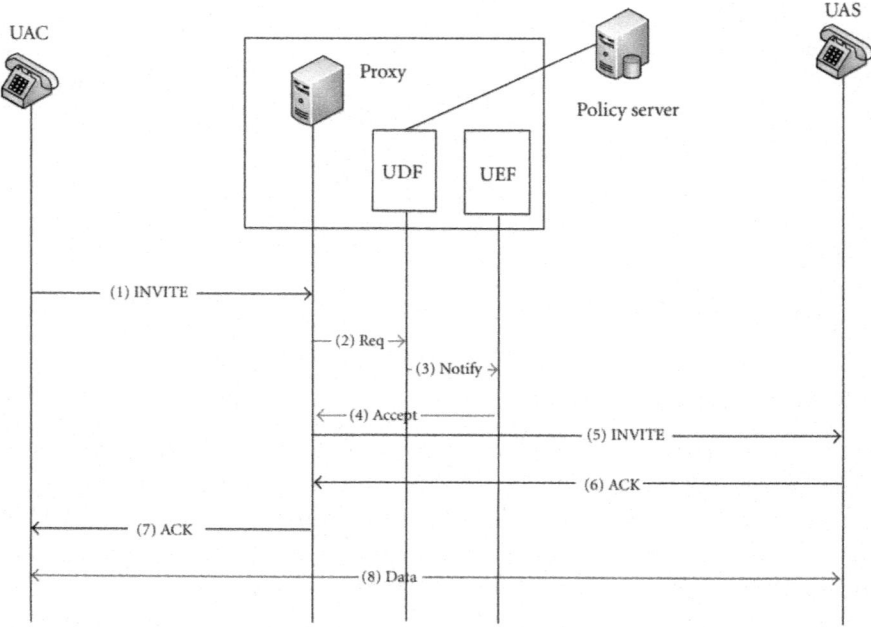

Figure 6: Authorization (before).

5.   The proxy processes the call initiation and forwards the INVITE to UAS.

6.   If UAS accepts the call, it will send an ACK to the proxy.

7.   The proxy relays the UAS ACK to the UAC and includes within the message the details of UAS (contact address, call attributes, etc.)

8.   At this point, the voice channel is established and UAC and UAS can now start communicating.

Application Scenarios

Scenario 1: UCON allows SIP to continuously monitor and control a call session even after it is established. In the case of pay-per-usage, UCON may terminate the call when the caller runs out of credit. This is done through the UDF and UEF. After the caller's credit has been validated, the UDF at the SIP server sends the details of the credit to the UEF at the client side, embedded within the parameters of ACK message. The UEF uses the update module to keep track of the credit consumed. When the credit reached its limit, the UEF terminates the call.

Scenario 2: System abuse: UCON allows SIP clients to control who can call them and when. This is done through the use of authorization before process (described in Figure 7). This process relies on the UDF and UEF to decide on whether a UAC is permitted to contact a UAS at any given time. To accomplish this task, access rules can be dynamically updated on the policy server to allow the ADF to look up these rules any time a call request is made and makes a decision on whether UAC is authorized to make call to UAS, and sends locally a notification to the AEF for the enforcement of such decision. In this case, the decision to allow a call to be made depends upon the positive acknowledgment that the proxy relays from the UAS to the UAC and includes within the message the details of UAS (contact address, call attributes, etc.).

Scenario 3: UCON allows SIP to set some obligations that users agree to perform prior to establishing a call. For example, a user may agree to log certain type of information pertaining to the call that he/she is requesting. Through the obligation module, within the ADF, SIP can terminate the call if that obligation was not fulfilled.

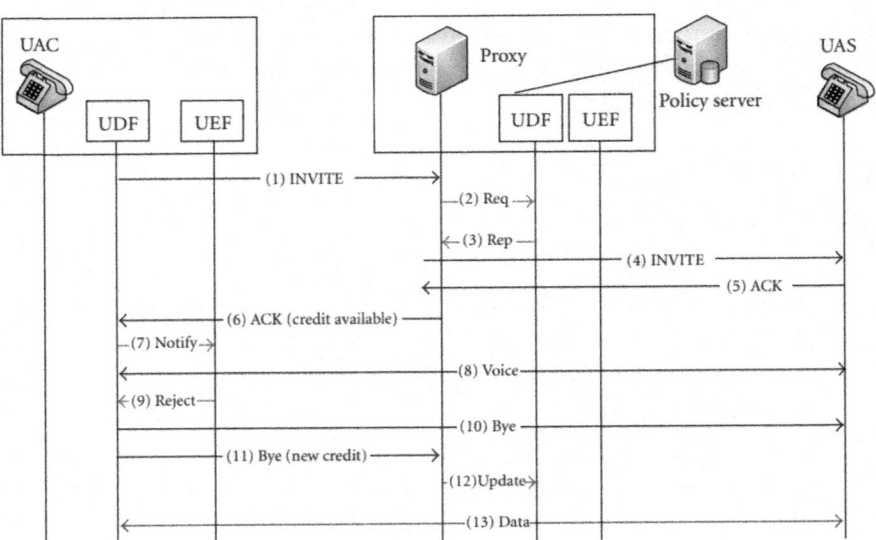

Figure 7: Authorization (during).

CONCLUSION

In this paper, we introduced a new concept of access control for enhancing the security of the SIP-based communications. This concept is based on controlling usage of the various SIP components. We have integrated this new concept with the SIP protocol to produce new architecture that helps in controlling the access to the SIP-based environment before, during, and after connections. Our architecture goes beyond what traditional access control models can provide. UCON benefits to SIP by providing comprehensive, systematic approach to fulfill the security requirements of today's pervasive and ubiquitous nature. In this work, we have presented an extended architecture which integrates UCON with SIP components in a manner that enables SIP to support applications in ubiquitous environment more securely. We have presented few scenarios where we showed the necessary mapping between UCON components and that of SIP. It is worth noting that this work provides mainly an architecture for applying UCON features to SIP-based communications in the context of usage and access control. However, this work does not pretend to provide a set of security mechanisms such as those for authorization and authentication algorithms.

This work opens several directions for further investigation. First of all, this architecture requires validation through a real-life implementation and deployment. Secondly, for this model to be complete a performance analysis of the integrated model is needed. We intend to extend this work in the future by providing an implementation of this architecture, and assess its performance with regards to its benefits.

REFERENCES

1.   J. Rosenberg, H. Schulzrinne, G. Camarillo, et al., "SIP: Session Initiation Protocol," Internet Engineering Task Force, RFC3261, June 2002.

2.   3GPP, "TS 24.228: signaling flows for the IP multimedia call control based on SIP and SDP," September 2002.

3.   IEEE std. 802.11, Wireless Medium Access Control and Physical Layer Specifications, 1999.

4.   R. S. Sandhu, E. J. Coyne, H. L. Feinstein, and C. E. Youman, "Role-based access control models,"Computer, vol. 29, no. 2, pp. 38–47, 1996. View at Publisher · View at Google Scholar

5.   W. Zhou and C. Meinel, "Implement Role-Based Access Control with Attribute Certificates," Universität Trier, Mathematik-Informatik, Forschungsbericht 03-03, 2003.

6.   W. Stallings, Network Security Essentials, Applications and Standards, Prentice Hall, Upper Saddle River, NJ, USA, 2003.

7.   R. Thomas and R. S. Sandhu, Task-Based Authorization Control (TBAC): Models for Active and Enterprise-Oriented Authorization Management

Database Security XI: Status and Prospects, North-Holland, Amsterdam, The Netherlands, 1997.

8. J. Park and R. S. Sandhu, "The UCONABC usage control model," ACM Transaction on Information and System Security, vol. 7, no. 1, pp. 128–174, 2004. View at Publisher

9. J. Park and R. S. Sandhu, "Towards usage control models: beyond traditional access control," inProceedings of the 7th ACM Symposium on Access Control Models and Technologies (SACMAT ⟨02), pp. 57–64, Monterey, Calif, USA, June 2002.

10. X. Zhang, F. Parisi-Presicce, J. Park, and R. S. Sandhu, "Logical model and specification of usage control," to appear in ACM Transaction on Information and System Security.

11. ISO/IEC 10181-3, Information Technology—Open System Interconnection—Security framework for open systems: Access control, http://www.iso.org/.

Chapter 4

CODE SYNCHRONIZATION ALGORITHM BASED ON SEGMENT CORRELATION IN SPREAD SPECTRUM COMMUNICATION

Aohan Li[1], Ziheng Yang [1], Renji Qi[1], Feng Zhou[1] and Guangjie Han[2]

[1]School of Electrical Engineering, Heilongjiang University, Harbin 150800, China

[2]Department of Information and Communication System, Hohai University, Changzhou 213022, China

ABSTRACT

Spread Spectrum (SPSP) Communication is the theoretical basis of Direct Sequence Spread Spectrum (DSSS) transceiver technology. Spreading code, modulation, demodulation, carrier synchronization and code synchronization in SPSP communications are the core parts of DSSS transceivers. This paper focuses on the code synchronization problem in SPSP communications. A novel code synchronization algorithm based on segment correlation is proposed. The proposed algorithm can effectively deal with the informational misjudgment caused by the unreasonable data acquisition times. This misjudgment may lead to an inability of DSSS receivers to restore transmitted signals. Simulation results show the feasibility of a DSSS transceiver design based on the proposed code synchronization algorithm. Finally, the communication functions of the DSSS transceiver based on the proposed code synchronization algorithm are implemented on Field Programmable Gate Array (FPGA).

INTRODUCTION

Software radio is a frontier technology in the field of wireless communications. Software radio has many advantages, such as modularization, high flexibility and good expansibility. Hence, software radio has become a popular research area [1,2]. Software radio technology can be applied on hardware platforms, which can implement different functions through software loading [3,4].

The anti-interference and confidentiality properties of the Direct Sequence Spread Spectrum (DSSS) technique are good. In addition, the power consumption of DSSS is very low, and hence DSSS has become a widely applicable communication technology [5]. The core technologies of DSSS are spreading code, modulation, demodulation, carrier synchronization and code synchronization. The performance of the code synchronization algorithm will directly influence the performance of the transceiver [6,7,8]. There are several existing code synchronization algorithms and, in general, they can be classified into frequency domain synchronization and time domain synchronization. Frequency domain synchronization has more advantages than the time domain synchronization, Such as the fact that it is better suited for avoiding interference and, in particular, for avoiding interference caused by the multi-path effect. Besides, frequency domain synchronization algorithms provide better solutions to the problems caused by the Doppler Effect and the different frequencies between the receiver and the transmitter.

Code synchronization mainly consists of two parts, which are code acquisition and code tracking. During the code acquisition phase, the phase difference between the pseudo-random code generated by the digital receiver and the received pseudo-random code needs to be checked against a certain threshold value. The threshold value is usually set as the minimum chip width, which is less than half the pseudo-random code. The code acquisition method can be generally classified into serial acquisition, parallel acquisition, multi-channel correlation acquisition, matched filter code acquisition*etc.* However, traditional serial acquisition methods based on correlation do not perform will in code acquisition loops based on Binary Phase Shift Keying (BPSK) modulation and demodulation. Although these methods can perform code acquisition, their capture efficiency is too low. The main problem of traditional serial code acquisition methods can be summarized as below: Because of the unreasonable data acquisition time signals transmitted by DSSS transmitters may not be accurately restored by the corresponding DSSS receiver. The data acquisition time is defined as the time points at which the code acquisition module of DSSS receiver collects the data points of the received signal.

In order to deal with the problems of traditional serial code acquisition, and improve the performance of code acquisition in practical engineering applications, we improved on the traditional serial code acquisition method by proposing a novel serial code acquisition method based on segment correlation. When using our proposed serial code acquisition, DSSS receivers can restore the signals transmitted by DSSS transmitter accurately, both in the time and the frequency domain. The code acquisition method, which is based on segment correlation, is summarized as follows: We partition each acquisition time interval into shorter time periods. We then apply a correlation operation between the segmented data contained within each of these time periods. After the correlation operation, we convert the results into absolute values and add them. The accumulated value is considered to be the final code acquisition value of each acquisition time interval. This proposed code acquisition method effectively solves the informational misjudgment caused by the unreasonable data acquisition time.

This paper focuses on the implementation of the DSSS transceiver based on the proposed code acquisition algorithm. The main contributions of this paper are summarized as follows:

(1)   A novel serial code acquisition method based on segment correlation is proposed, which effectively solves the informational misjudgment caused by the unreasonable data acquisition time.

(2)   An overall communication DSSS transceiver system solution based on the proposed novel code acquisition is introduced.

(3)   The function of the different modules of the DSSS transceiver based on the proposed code synchronization algorithm was verified through MATLAB simulations. The tested modules include M-sequence, BPSK modulation, BPSK demodulation, Numerically Controlled Oscillator (NCO), low pass filter, carrier synchronization and code synchronization.

(4)   The verified modules were then simulated using Verilog language in order to load each module codes to a FPGA system and then verify the overall system's functionality. The general implementation of the DSSS transceiver can be summarized as below: the DSSS transmitter and receiver are both linked with a computer through a serial port line. The computer transmits signals through the digital transmitter using Com Wizard. Signals are transmitted to the digital receiver after SPSP and modulation. Then the received signals are restored through demodulation, carrier and code synchronization. The restored signals are conveyed to the computer, which is linked with the digital receiver. Hence, we can verify whether the original signal is restored by the computer. In this manner, the feasibility of our proposed code synchronization algorithm was confirmed through the implementation of the DSSS transceiver on FPGA.

(5)   For the purposes of this paper, a PCB layout was also designed and physically constructed.

(6)   The digital DSSS transceiver's behavior was also analyzed through the use of Signaltap II.

The rest of this paper is organized as follows: the overall communication solution is proposed in Section 2. In Section 3, we describe the code synchronization algorithm based on segment correlation. Modules simulation via MATLAB is presented in Section 4. In Section 5, we describe the implementation of the DSSS transceiver based on the proposed code synchronization algorithm on FPGA. In Section 5, we conclude this paper.

OVERALL COMMUNICATION SOLUTION OF DSSS TRANSCEIVER

This section describes several core modules of the DSSS transceiver, namely the spreading code module, the modulation and demodulation modules and the carrier synchronization module.

Spreading Code

M-sequence has an excellent autocorrelation property while it is also easy to generate, replicate and control. M-sequence also has a certain independent address number and a quite long sequence period. It is usually used as the spreading code in SPSP Communication [9]. M-sequence is the longest linear shifting register sequence, and is generated by a feedback shift register. The cycle of the m pseudorandom sequence generator used in this paper is 1023. Equation (1) is the polynomial of the M-sequence whose cycle is 1023. The M-sequence generator used in this paper is shown in Figure 1.

$$f(x) = 1 + x_7 + x_{10} \tag{1}$$

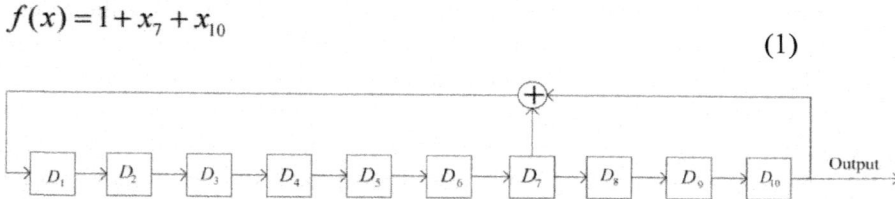

Figure 1: M-sequence generator.

Modulation and Demodulation

The BPSK modulation and demodulation technique is used in this paper. BPSK has some advantages, such as high spectrum efficiency, strong anti-interference properties, good spectral characteristics, quick transmissibility and others. BPSK uses the 0 s and 1 s of the digital signal to switch between a 0 and π phase of the carrier signal [10,11,12,13]. Equation (2) is the time-domain expression of the BPSK signal.

$$s(t) = \left[\sum_{n=-\infty}^{\infty} a_n g_T(t - nT_b) \right] \cos(\omega_i t + \theta_i) \tag{2}$$

where a_n denotes a bipolar sequence, whose statistic characteristic values are 1 and −1. w_i denotes the carrier frequency. T_b denotes the binary symbol interval. θ_i is the initial phase of the carrier. $g_T(t)$ denotes a single rectangle pulse, whose pulse width is T_b. $g_T(t)$ has raised cosine characteristics.

In this paper, as the selection method to generate the BPSK signal we used the baseband signal to control a switching circuit to select the input signal. The input signal of the switching circuit is two co-frequency carriers whose phase difference is ϖ. Figure 2 shows the block diagram of our BPSK modulation module:

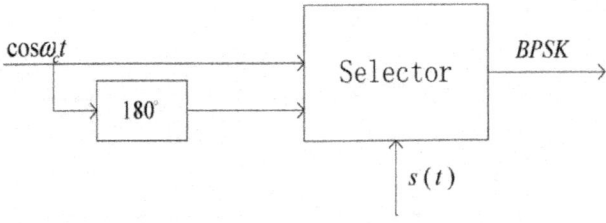

Figure 2: Block diagram of Binary Phase Shift Keying (BPSK) modulation.

Coherent demodulation and incoherent demodulation are two common demodulation methods. Coherent demodulation has great advantages with regard to threshold detection, output signal to noise ratio, error rate and others, and is the method used in this paper. The block diagram of the BPSK demodulation module is shown in Figure 3.

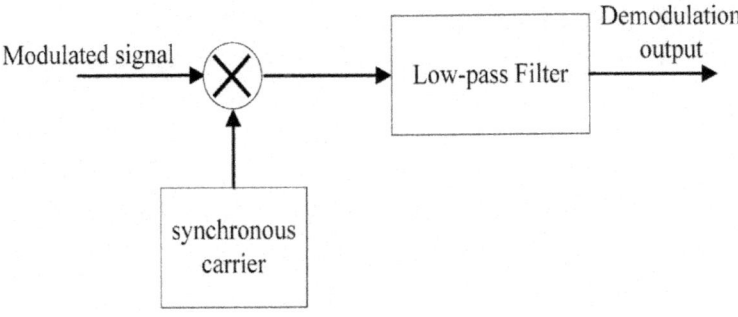

Figure 3: Block diagram of BPSK demodulation.

Carrier Synchronization

The Costas loop synchronization method is used in the carrier synchronization module [14]. Coatas loop synchronization belongs to the direct methods [15]. Compared to the pilot carrier method, direct method can extract the carrier from the signals without a pilot signal. Direct methods can achieve carrier synchronization using the modulated signal as an input. Direct methods not only save resources but are also very useful in cases of limited spectrum resources [16]. The block diagram of the Costas loop is shown in Figure 4. The process of the Costas loop can be expressed mathematically as follows:

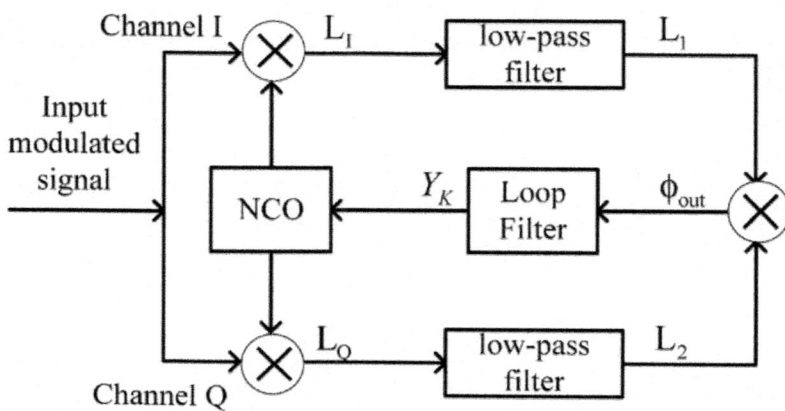

Figure 4: Block diagram of the Costas loop.

Let us define the input signal of the Costas loop as $m(t)\cos\phi t$, where $m(t)$ denotes the baseband signal, and ϕ denotes the angular frequency of the carrier of the transmitter. Hence, the signals of the orthorhombic channel I and channel Q can be expressed as Equations (3) and (4), respectively:

$$I(t) = \cos(\phi t + \theta) \tag{3}$$

$$Q(t) = \sin(\phi t + \theta) \tag{4}$$

The phase difference between the local carrier signal and the input modulated signal is φ. Hence, we obtain Equations (5) and (6).

$$L_I(t) = m(t)\cos\phi t \times \cos(\phi t + \varphi) = \frac{1}{2}m(t)\left[\cos\varphi + \cos(2\phi t + \varphi)\right] \tag{5}$$

$$L_Q(t) = m(t)\cos\phi t \times \sin(\phi t + \varphi) = \frac{1}{2}m(t)\left[\sin\varphi + \sin(2\phi t + \varphi)\right] \tag{6}$$

Equations (7) and (8) are obtained after $L_I(t)$ and $L_Q(t)$ go through the low-pass filter:

$$L_1(t) = \frac{1}{2}m(t)\cos\varphi \tag{7}$$

$$L_2(t) = \frac{1}{2}m(t)\sin\varphi \tag{8}$$

We then multiply $L_1(t)$ by $L_2(t)$ to find the error signal, which is expressed by Equation (9):

$$\varphi_{out}(t) = L_1 L_2 \approx \frac{1}{4} m^2(t)\varphi$$

$$(9)$$

From the above equations, we see that the local NCO control signal will control the frequency and phase of NCO until the carrier loop is in synchronization [17]. The carrier signal can then be extracted by tracking the frequency and phase of the transmitter.

CODE SYNCHRONIZATION ALGORITHM BASED ON THE SEGMENT CORRELATION

First, the overall design scheme and block diagram of the code synchronization module are described in this section. Then code acquisition is described in detail, while existing problems of traditional series code acquisition methods are pointed out and corrected, and a novel series code acquisition method based on the segment correlation is implemented. Finally, the code tracking module is described.

Overall Design of the Code Synchronization Module

The overall block diagram of the code synchronization module in the digital transceiver is shown in Figure 5. The workflow of the code synchronization process can be summarized as follows:

(1)   A correlation operation between the received signal and the carrier signal is applied by the correlator.

(2)   Then, the computed value is analyzed by the code acquisition module. The phase difference between the carrier signal generated by the local NCO and the carrier signal of the received signal will meet the design requirement after it passes through the code acquisition module.

(3)   Tracking operation and adjustment will be performed. The phase difference between the local carrier signal and the carrier of the received signal is further decreased during the tracking process by adjusting the clock of the local code generator step by step [18].

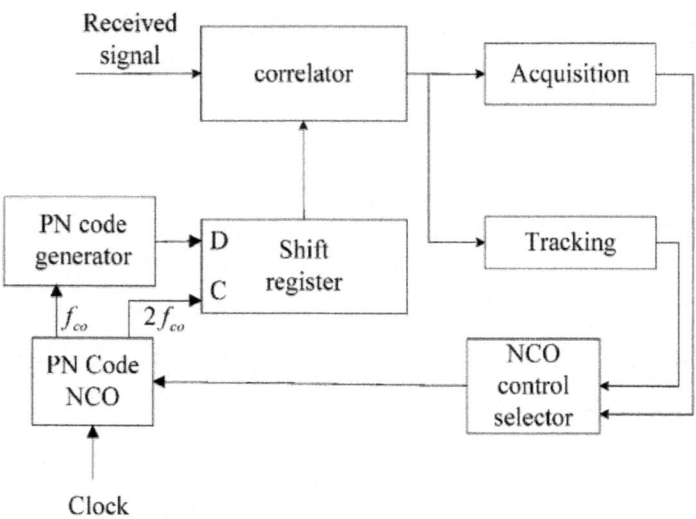

Figure 5: Block diagram of the code synchronization module.

The code acquisition module and the code tracking module are the two main components of the code synchronization module. The main work of our paper is to propose a novel code acquisition algorithm.

Design of the Code Acquisition Module

The block diagram of the acquisition loop of the receiver is shown in Figure 6. The acquisition judgment circuit is the core component of the acquisition loop. The serial acquisition method is used in the acquisition judgment module. The serial acquisition method is well used in situations where the cycle of the carrier signal is short. However, some problems occur when the cycle of the carrier signal is too long. Specific analyses of various such scenarios are described as follows:

Situation 1: the carrier signal is not inverted after BPSK demodulation. In our case, the word "inverted" means that positive carrier signals before BPSK demodulation turn into negative carrier signals after BPSK demodulation. The signal waveform of such a situation is shown in Figure 7. After the correlation accumulation operation, the output signal is positive peak-to-peak. Acquisition judgment methods can succeed in such cases, *i.e.* carrier signals can be correctly restored after BPSK demodulation.

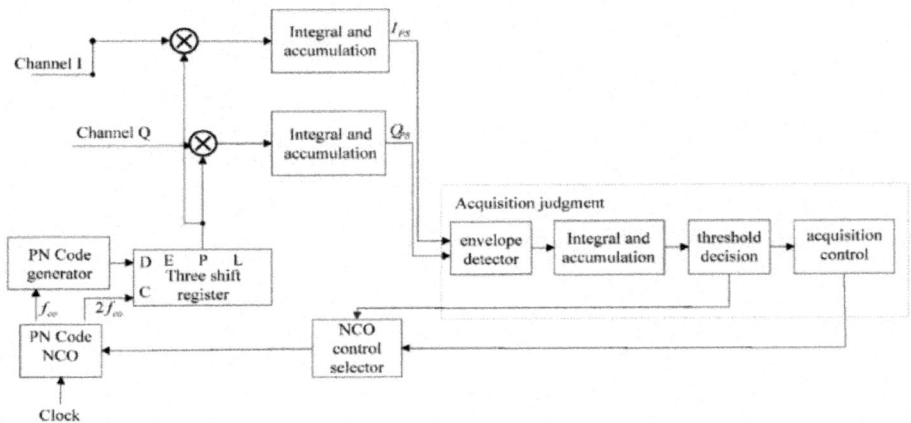

Figure 6: The block diagram of the acquisition loop of the receiver.

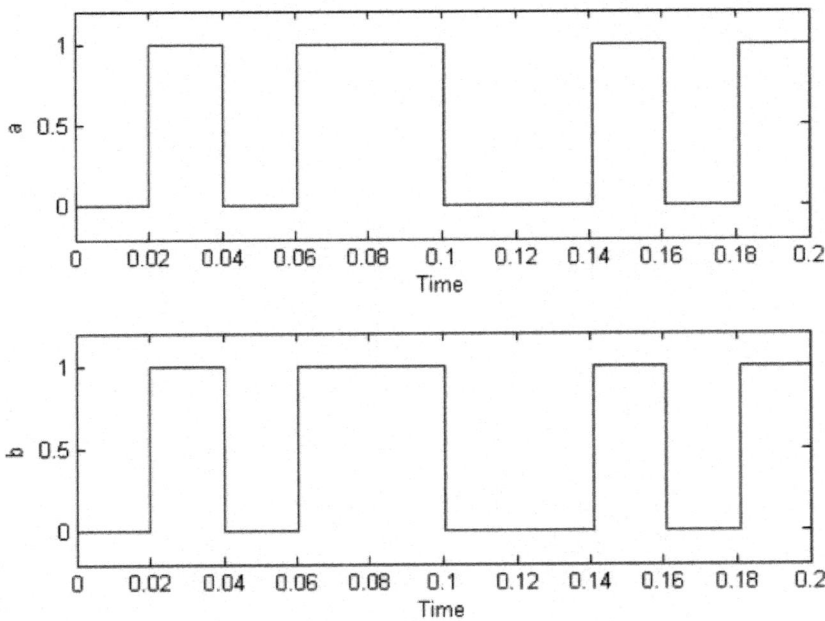

Figure 7: Carrier signal example indicative of the first situation.

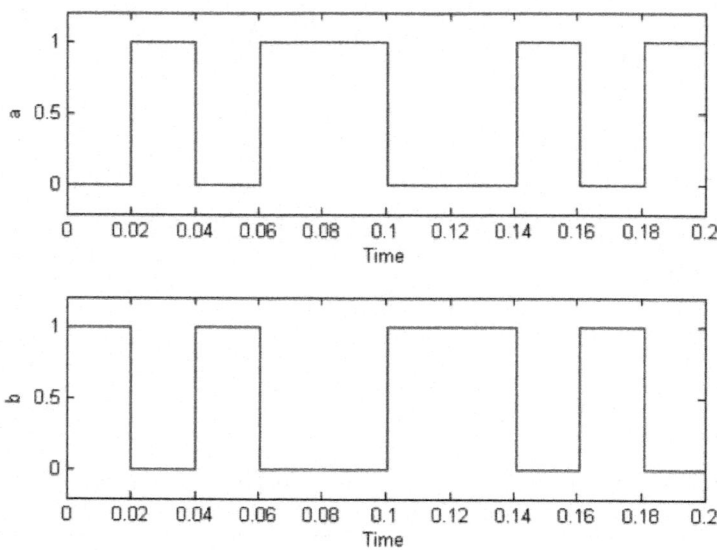

Figure 8: Carrier signal example indicative of the second situation.

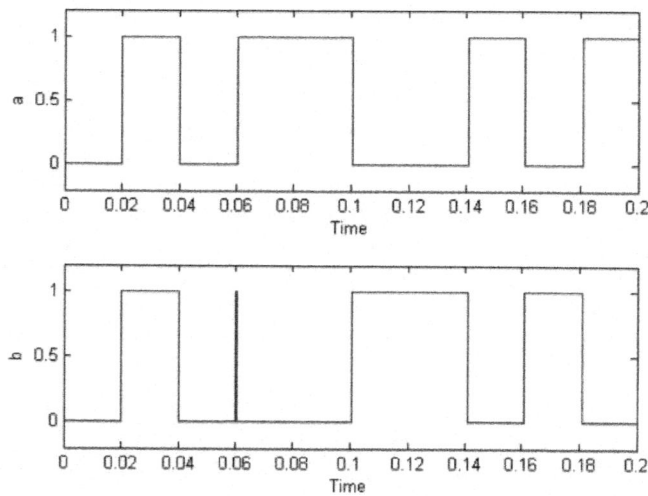

Figure 9: Carrier Signal of the third situation.

Situation 2: the carrier signal is inverted after BPSK demodulation, while the data acquisition time of the correlative accumulation is reasonable. A sample signal waveform of such a situation is shown in Figure 8. After the correlation accumulation operation, the output signal is negative peak-to-peak.

Acquisition judgments also can succeed in such situations.

Situation 3: the carrier signal is inverted after BPSK demodulation, but the data acquisition time of the correlative accumulation is not reasonable. This situation is shown in Figure 9. In this situation, although the signal is acquired, the value after accumulation is far less than the threshold value which we have set and erroneous judgment occurs.

Hence, to avoid unsuccessful acquisitions, a novel serial acquisition algorithm based on segment correlation is proposed.

Unlike traditional code acquisition algorithms, our proposed acquisition algorithm is based on correlation that is performed after partitioning each acquisition time interval into smaller segments. Then, a correlation operation is performed between acquisition data for each such segment. After the correlation operation, the absolute values of the obtained correlation operation results of segment are summed. The accumulated value is used as the final code acquisition value of each acquisition time interval. The block diagram of the proposed serial acquisition algorithm is shown in Figure 10. In the proposed serial acquisition algorithm, the absolute value of the results obtained after the correlation operation are the peak-to-peak values in each time segment. Peak-to-peak values will be acquired by adding the peak-to-peak values in each segment of the acquisition time interval. In this manner, erroneous judgments are avoided and, hence, the capturing efficiency is improved.

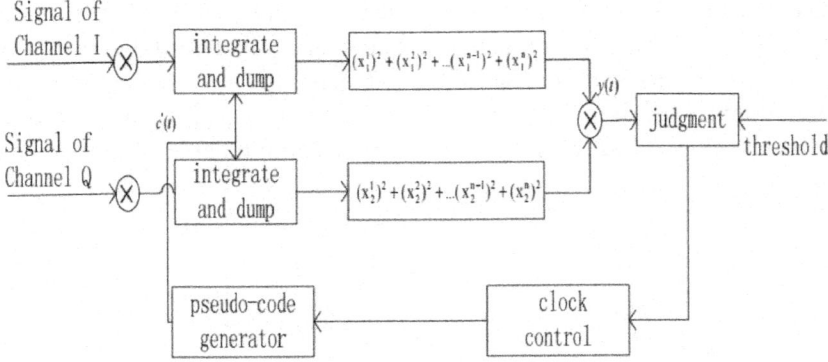

Figure 10: Block diagram of the proposed serial acquisition algorithm.

Code Tracking

The block diagram of code tracking loop module used in this paper, which is based on a delay-locked loop, is shown inFigure 11. Its main components are the code tracking loop discriminator and loop filter.

The algorithm of the discriminator can be expressed through Equation 10:

$$\frac{1}{2}\frac{E-L}{E+L}$$

$$(10)$$

where, $E = \sqrt{I_{ES}^2 + Q_{ES}^2}$ and $L = \sqrt{I_{LS}^2 + Q_{LS}^2}$.

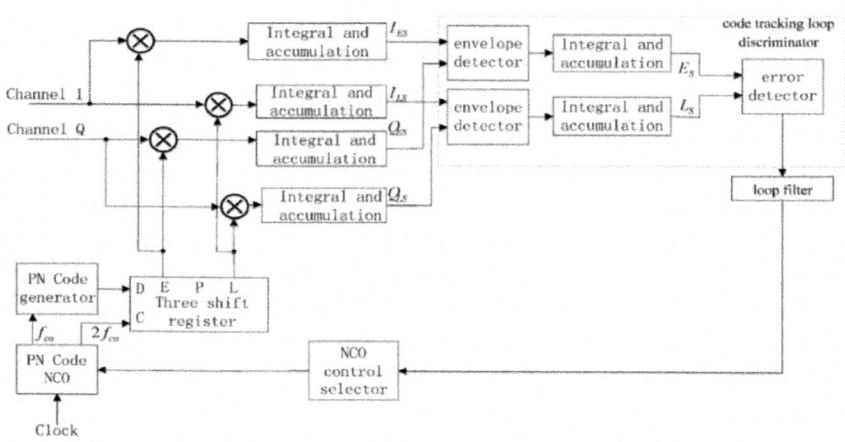

Figure 11: Block diagram of code tracking loop module.

The loop filter plays an important role in the code tracking loop. Compared with a first-order or a third-order loop filter, a second-order loop filter has many advantages, such as: the frequency deviation is constant; the zero-frequency gain is infinite; its stability is very good it is also easy to implement. This paper adopts a second-order loop filter solution, the frequency domain model of which is shown in Figure 12. z^{-1} denotes the unit delay of the integrator.

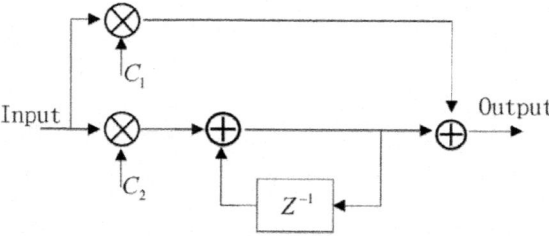

Figure 12: The frequency domain model of the second-order digital loop filter.

SIMULATION

In this section, the simulations of the M-sequence, the BPSK modulation, the NCO module, the low-pass filter, the carrier synchronization, and the code synchronization modules of the DSSS transceiver, based on the proposed code synchronization algorithm, are presented, in order to demonstrate the role of each of the different modules and prove the advantages of our approach.

M-Sequence Module

An M-sequence is first generated by a shift register. The Matlab-generated simulation result of the M-sequence module is shown in Figure 13.

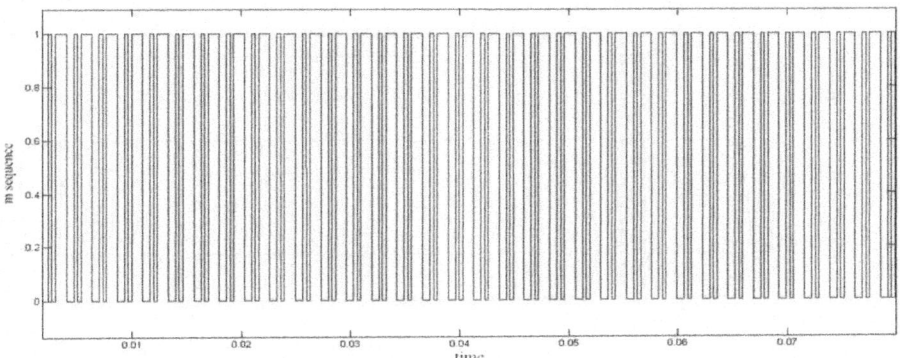

Figure 13: The simulation result of the M-sequence module.

BPSK Modulation

We then multiply a baseband signal by the M-sequence to get SPSP. The simulation result, as obtained using Matlab, is shown in Figure 14.

The SPSP sequence will be obtained after the baseband signal has been spread. BPSK modulation is then performed by passing the SPSP sequence through a modulator. The simulation result of the BPSK modulation is shown in Figure 15.

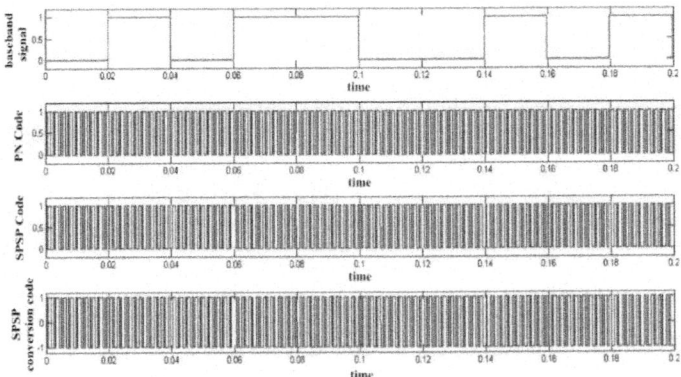

Figure 14: Simulation result of Spread Spectrum (SPSP).

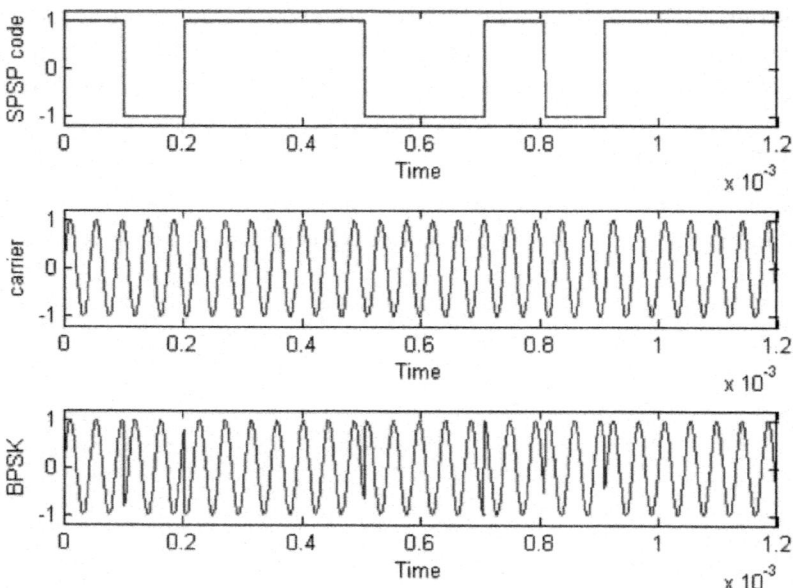

Figure 15: The simulation result of BPSK modulation.

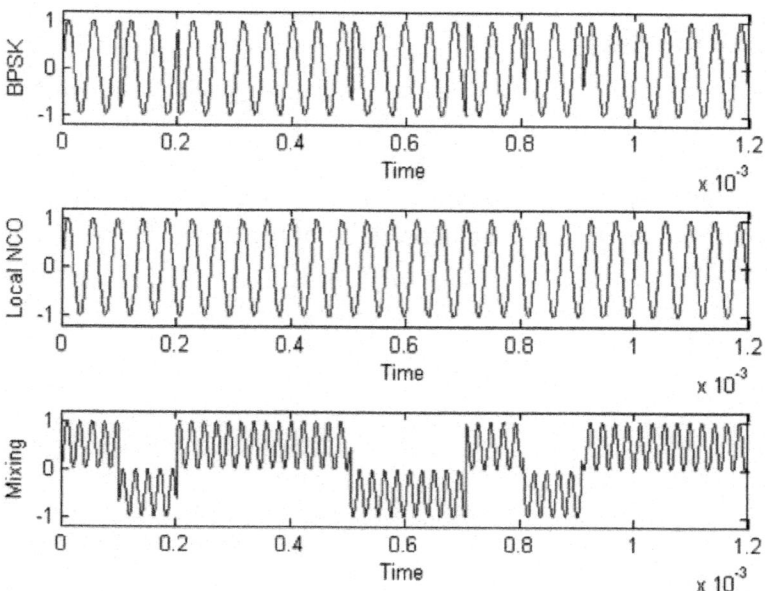

Figure 16: Simulation result of demodulated waveform.

After spreading the spectrum and modulating the baseband signal, we transmit it and demodulate the waveform to verify whether we can recover the original SPSP waveform correctly. The demodulated waveform is shown in Figure 16, from which we see that the demodulated waveform is as same as our initial one shown in Figure 14.

NCO Module

The purpose of the NCO module is to generate two mutually orthogonal signals. The technological basis of NCO is Direct Digital Synthesizer (DDS) technology [19,20,21]. The simulation result of the NCO module output is shown in Figure 17.

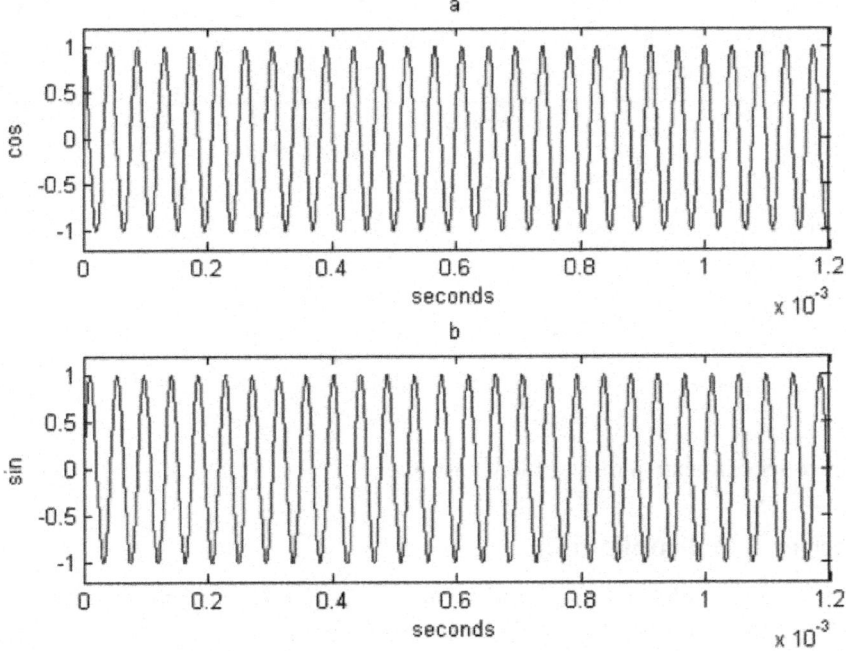

Figure 17: Simulation result of NCO.

Low-Pass Filter

From Figure 17, we can see that there are many harmonic components, which do not meet our design requirements. The signals that we want to obtain are zero medium frequency signals. Hence, we need to process the mixed signal effectively, which implies using a low-pass filter. The low-pass filter in this paper is implemented through an integration and accumulation manner. The main function of the low-pass filter is to filter high frequency components. The simulation result of the low-pass filter for the mixing output of Figure 16 is shown in Figure 18.

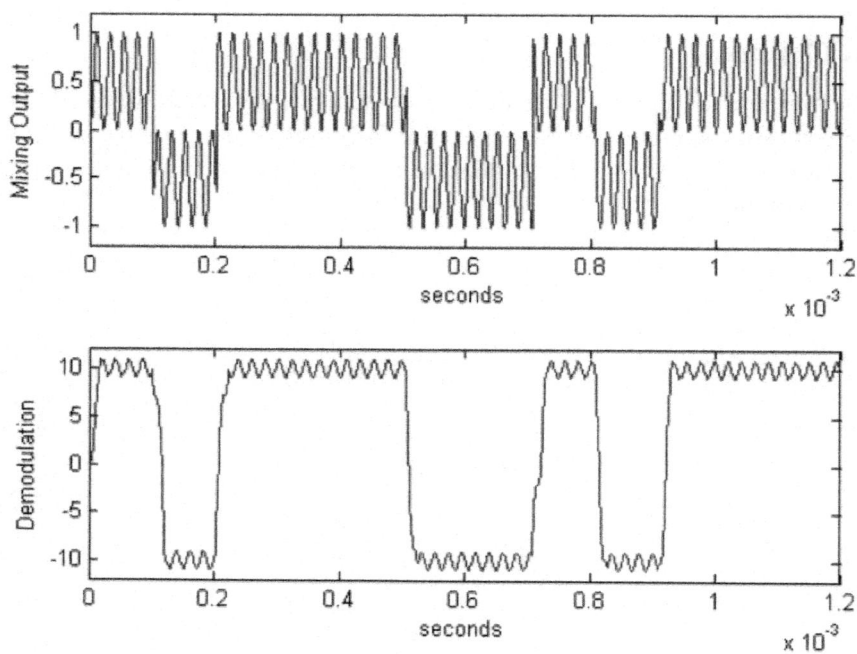

Figure 18: Simulation result of low-pass filter.

Carrier Synchronization

The carrier synchronization module used in our system is based on the Costas loop. In this module, the carrier loop phase detector constitutes an important component, as it is used to compare phases [22]. This paper uses CORDIC algorithm to implement an arc tangent phase detector [23,24]. The simulation result of the phase detector is shown in Figure 19, from which we can see that at the beginning of loop, a phase difference exists, but the output value of the phase detector will tend towards 0 as the carrier approaches synchronization.

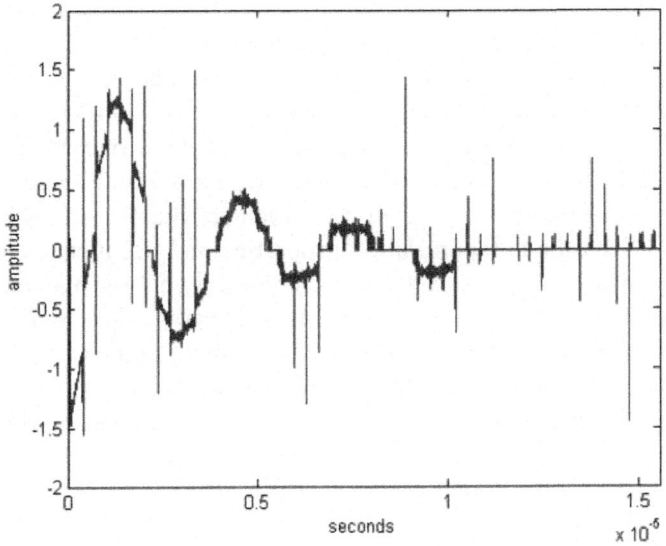

Figure 19: Simulation result of phase detector.

The simulation result of the carrier loop is shown in Figure 20. In this figure, we can see that high frequency components have been filtered and that the baseband signal will be obtained after the received signal is demodulated.

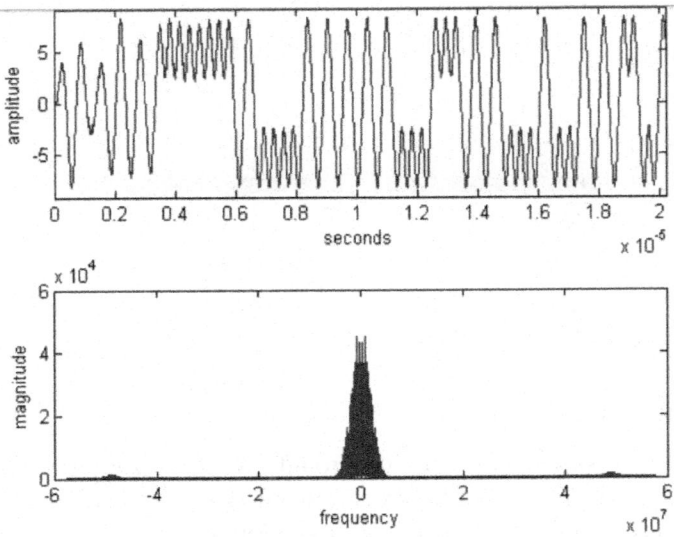

Figure 20: Simulation result of carrier loop.

Code Acquisition

The purpose of the code acquisition simulation is to verify the theoretical feasibility of code acquisition loop. In this simulation, two M-sequences whose phases are different are first generated. These two M-sequences have same cycle. Following the circular method, we slide one M-sequence as a correlative operation is conducted among the points in one fixed cycle during slide process. Then, we sum the values which are the result of the correlative operation. Given the autocorrelation properties of M-sequences, peak values will appear when the phase difference of these two M-sequences meets the code acquisition requirement; otherwise, near-zero values will appear. The simulation result of code acquisition is shown in Figure 21, which shows peak values when the phase difference between two M-sequences meets the code acquisition requirement. Otherwise, the output values remain at very low levels.

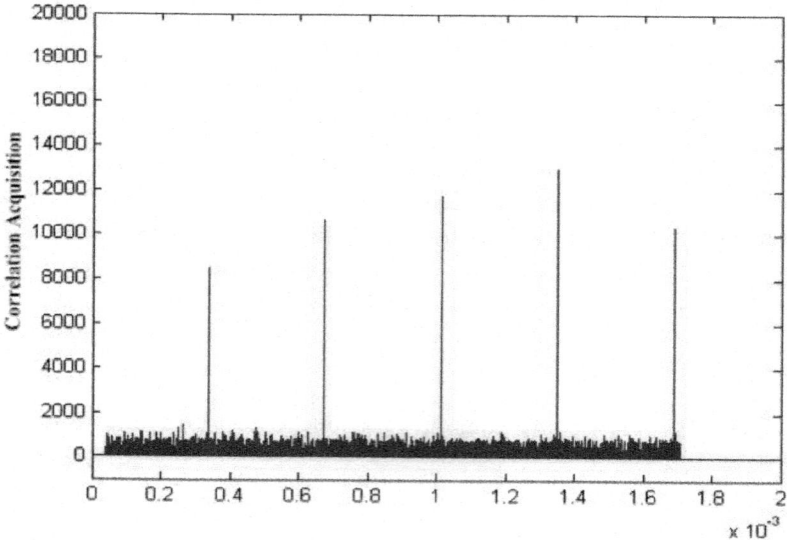

Figure 21: Simulation result of code acquisition.

Code Tracking

The code acquisition loop provides rough synchronization while the code tracking module gives us subtle synchronization. Code tracking cannot change the timer of the code generator significantly, but does so step by step according to the effect of the phase detector and the loop filter to achieve the end goal [25]. The simulation result of the code tracking module is shown in Figure 22.

From Figure 22, we can see that the phase difference decreases with successive loop iterations and slowly converges to zero, which illustrates that the phase difference between two sequences are decreasing step-wise. The output of the loop filter changes with the output of phase difference. Besides, we can see that the output of loop filter is comparatively high when the phase difference is comparative big, which leads to the significant adjustment of the overall loop. Accordingly, the output of the loop filter is comparatively small when the phase difference small, and is equal to zero when there is no phase difference.

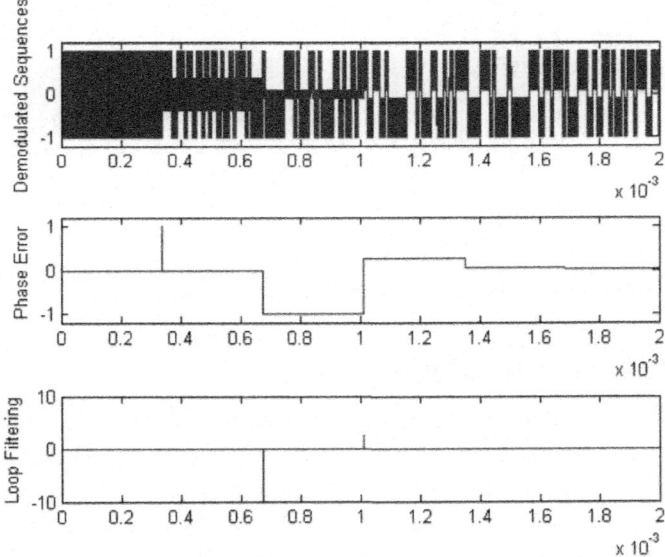

Figure 22: Simulation result of code tracking.

Code Synchronization

The simulation of overall code synchronization combines code acquisition and code tracking, demodulation of the modulated baseband signal through code synchronization, and restoration of the baseband signal. The simulation result of overall code synchronization is shown in Figure 23, which shows that the demodulated baseband signal is chaotic when the signal is not synchronized and correctly restored when the signal is synchronized. Hence, we see that the DSSS receiver restores the signals transmitted by the transmitter accurately with regard to both time and frequency, based on our proposed serial code acquisition.

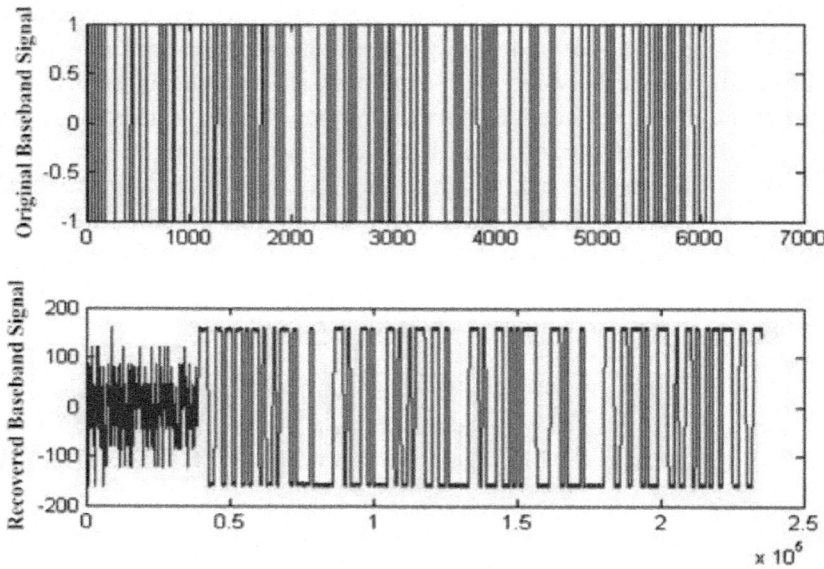

Figure 23: Simulation of code synchronization.

FPGA IMPLEMENTATION

The PCB design and manufacturing of the hardware platform of the proposed DSSS software transceiver are described in this section. The specific FPGA implementation of the carrier synchronization module, the code acquisition module, the code tracking module are also presented here, along with the implementation of the transmitter and receiver. The full software-based DSSS transceiver design was thus hardware tested on this platform, and its correctness and feasibility were verified through the demonstration of a communications' channel between two PCs.

PCB Design and Manufacture of the DSSS Transceiver System

Altium Designer software was used to design PCB in this paper. The main circuits of the DSSS transceiver consist of the following: FPGA core, clock, power, reset, A/D and D/A. A schematic was designed based on these circuits, and then converted to a PCB diagram. Our PCB design uses a double-layer structure and can meet high speed requirements. In addition, it was designed for the maximal suppression of electromagnetic compatibility and crosstalk. Finally, the designed PCB board is welded. The Physical map of the hardware platform is shown in Figure 24.

Figure 24: The Physical map of the hardware platform.

FPGA Implementation of the Carrier Synchronization Module

The filter, module, phase discrimination, integral accumulation and NCO modules are the main modules of the carrier loop in our FPGA implementation. The overall FPGA framework of the carrier synchronization module is shown in Figure 25. After integrating and adjusting the above modules, we used the logic analyzer of Quartus II for debugging. The signal waveform obtained from the logic analyzer is shown in Figure 26. From fil_cos and fil_sin in Figure 26, we see that the baseband signal has been restored, which shows that the FPGA implementation of the carrier synchronization module was for the most part successful.

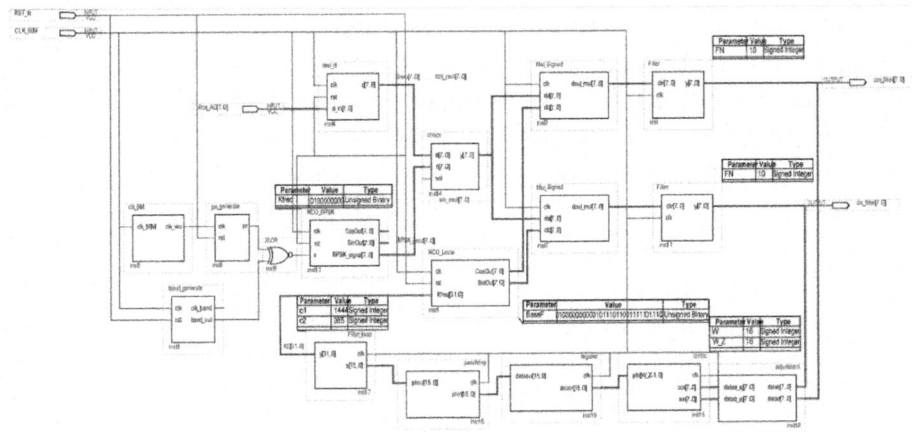

Figure 25: The overall FPGA framework of the carrier synchronization module.

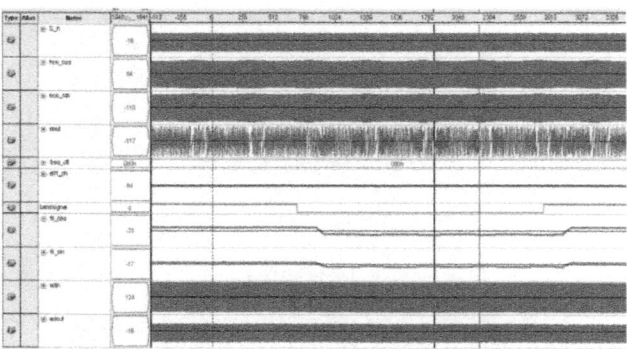

Figure 26: The waveform of carrier loop synchronization.

FPGA Implementation of the Code Acquisition Module

The FPGA implementation of the code acquisition module is shown in Figure 27. The module consists of the correlator, operation and judgment modules. The yihuo_mult module in Figure 27 is the correlator module. The main function of correlator module is to multiply local PN code by received signal. The integer_ capture module in Figure 27 is the operation module. The functions of the operation module are summarized as follows:

(1)   Segmental integration of the output values of the correlator module.

(2)   Acquisition of the absolute values of the segmental integrated values.

(3)   Summation of all the absolute values in one cycle.

(4)   Threshold judgment of the final added values to judge whether the k and the s signal have been acquired.

Judge module in Figure 27 is the judgment module. The main function of the judgment module is to judge whether the code acquisition is successful given the values of the k and s signals. G_out = 1 when the code acquisition is successful, in which case the code tracking loop can be initialized.

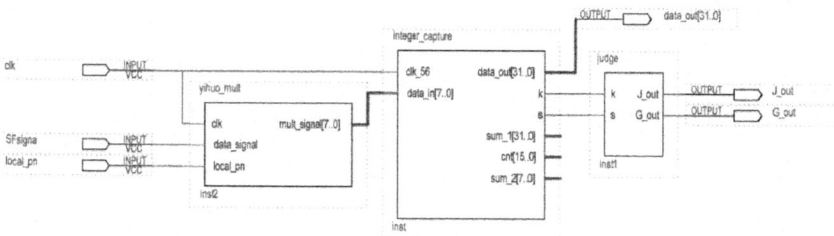

Figure 27: The top-level design of the code acquisition module.

After creating the FPGA code of the code acquisition module, a timing simulation of the programmed FPGA circuit was conducted. The simulation result is shown in Figure 28. The m_out signal in Figure 28 is the local PN sequence. m_out3 is the simulative received PN code sequences. k is the flag of unsuccessful acquisition. s is the flag of successful acquisition. When m_out and m_out3 are show that acquisition has been unsuccessful, the output of k is a high level signal while the output of s is a low level signal after one computing cycle. When m_out and m_out3 are acquired in phase, the output of k is a low level signal while the output of s is a high level signal after one computing cycle. These results confirm the proper function of our FPGA design for the code acquisition module.

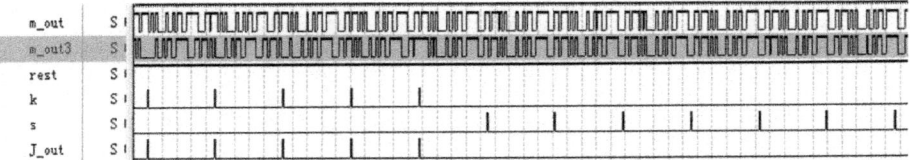

Figure 28: Timing simulation of the code acquisition module.

After the timing simulation, the code acquisition module was debugged with the use of SignalTap II. The debugging result is shown in Figure 29. m2 is the local PN code signal generated by the receiver. cos_out is the received PN code signal. G_out16 is the successful flag of code acquisition. From Figure 29, we can see that the output of G_out16 is at the high level when the phase difference between the local PN code signal and the received PN code signal is less than the minimum chip width.

Figure 29: The waveform of the code acquisition module.

From the figure, we also see that the proposed improved code acquisition loop system implements the function of the code acquisition module successfully.

FPGA Implementation of the Code Tracking Module

The FPGA module of the code tracking loop is shown in Figure 30. The main code tracking loop contains the module for the absolute value operation, along with the correlator, integrator, error detector and the loop filter modules.

yihuo_mult inFigure 30 is the correlator module, integer_track is the integrator module, jueduizhi is the operation module for obtaining the absolute values, while error_detect is the error detection module. The main function of the error detector module is to calculate the corresponding phase difference according to the correlation cumulative sum of the anticipated code and lag code between received code and local PN code. In addition, codeloopfilter in Figure 30 is the loop filter module. The function of this module is to control and adjust the clock of the local PN code generator according to the output error value of the error detection module.

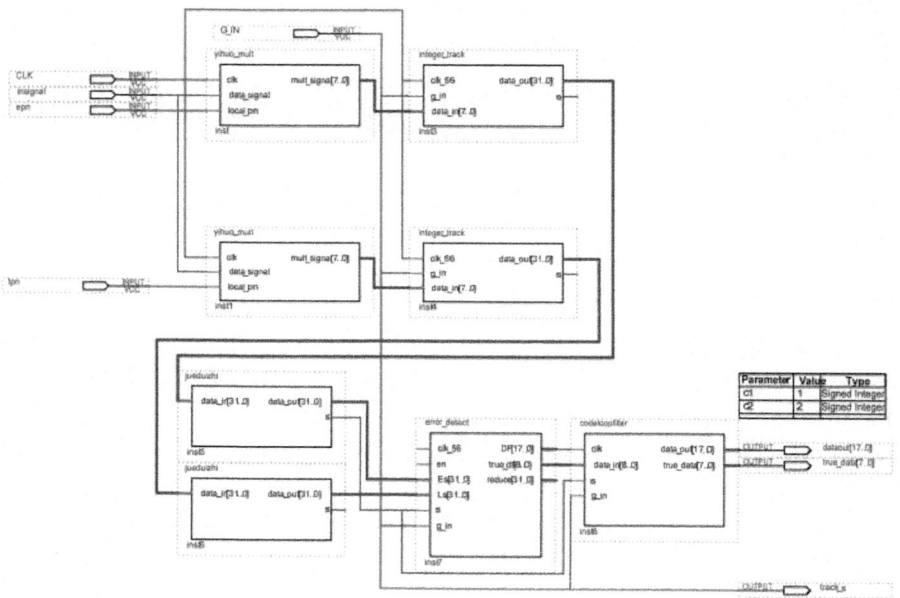

Figure 30: The top-level design of the code tracking module.

After Verilog coding, a timing simulation of the code tracking module was conducted, the results of which are shown inFigure 31. m1 and m2 are two PN code signals which have different phase. DF is the output value of the loop filter. Code tracking is successful when the error of the phase difference between the two signal channels is very small. At that moment, the output value of the loop filter is zero.

The result of the system debugging of the code tracking module is shown in Figure 32. filter_out is the synchronized version of the received signal. cos_ out signal is the adjusted filter_out signal, while m2 is the local PN code signal. filter_loop signal is the output value of the loop filter. From Figure 32, we can see that the output of the loop filter is zero when the local PN code signal tracks the received PN code signal, which verifies the design of the code tracking module.

In summary, the code synchronization system meets the requirements of the digital transceiver.

Figure 31: Timing simulation result of the code tracking module.

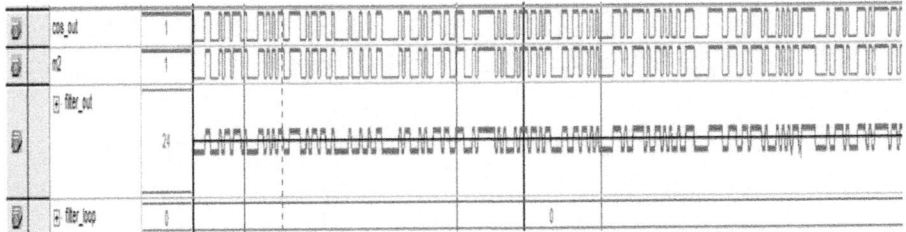

Figure 32: The waveform of the code tracking module.

FPGA Implementation of Transmitter

The top-level FPGA design of the DSSS transmitter is shown in Figure 33. It mainly contains four modules, which are PLL 280, kuopin, NCO BPSK and DAC out. PLL 280 is the SPSP module and is implemented by the SPSP function of Quartus II. The main function of PLL 280 is to generate a systematic clock according to the real system's requirements. kuopin is the BPSK SPSP module. The main function of kuopin is to make the baseband signal correlate with the local PN code, and then obtain the necessary SPSP code. NCO BPSK is the modulation module, the main function of which is to complete BPSK demodulation. DAC out is the driver module of D/A. The main function of DAC out is to send the modulated digital signal to the D/A converter.

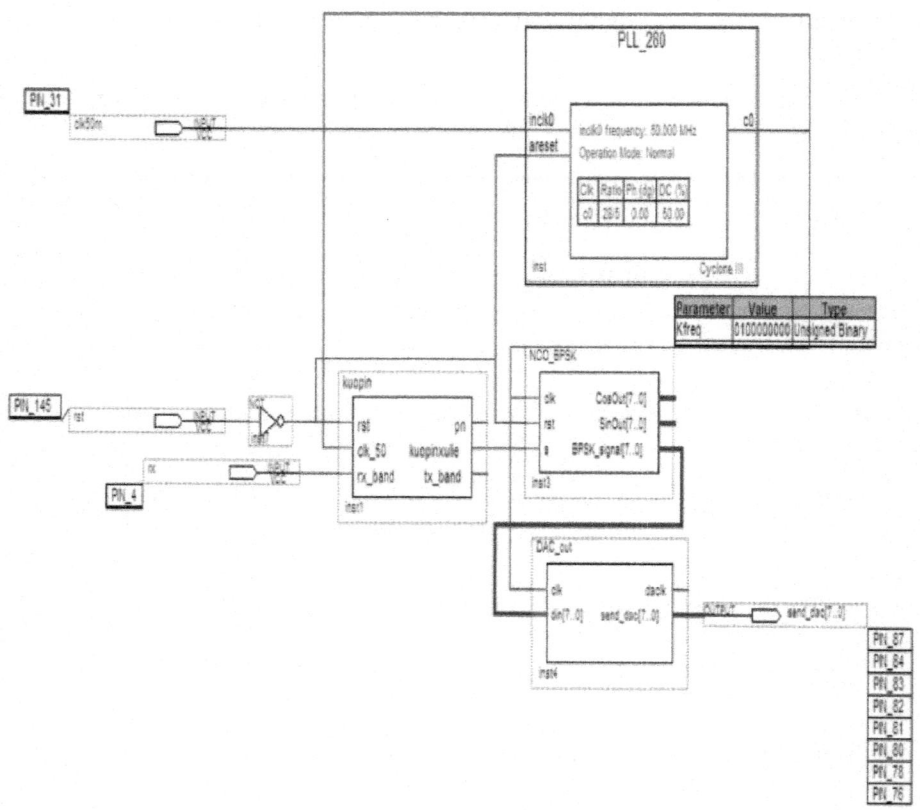

Figure 33: The top-level design of the Direct Sequence Spread Spectrum (DSSS) transmitter.

FPGA Implementation of Receiver

The receiver system is to implement the carrier synchronization loop and the code synchronization loop. The FPGA top-level design of the DSSS receiver is shown in Figure 34. From this figure, we can see that FPGA receiver contains AD Ctrl, SDRrec Carrytongbu, Capture, Track and Txuart.

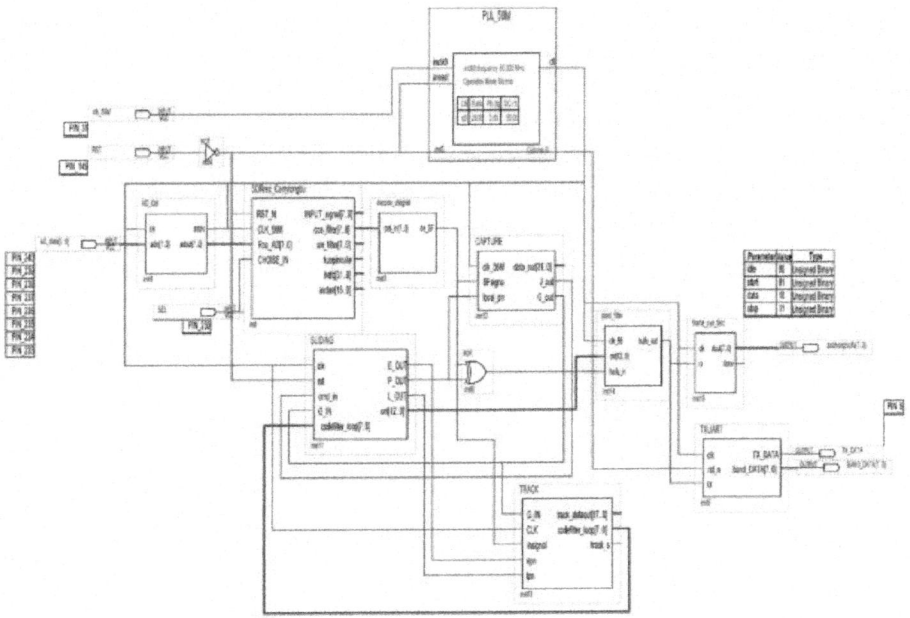

Figure 34: The top-level design of DSSS receiver.

AD Ctrl is the driver module of the A/D chip. The main function of AD Ctrl is to control the A/D chip, transform the received analog signal into a digital signal, and then load the digital signal to the FPGA. SDRrec Carrytongbu is the carrier synchronization module. Its main function is to load the programmed carrier codes to the FPGA platform. CAPTURE and TRACK are the code acquisition module and code tracking module, respectively. The function of CAPTURE and TRACK are to load the programmed code synchronization module to the FPGA platform. TXUART is the serial port driver module and its function is to transmit the restored original baseband signal to the PC through the serial port. The transmitted information can then be analyzed using PC software.

Overall Experiment and Demonstration of DSSS Transceiver

First, the DSSS transmitter and receiver are linked by coaxial-cable. Then, the DSSS transmitter and receiver are each linked with a PC using a serial port line. Using Com Wizard, the PC supplies information to the digital transmitter; this information is transmitted to the digital receiver after SPSP and modulation. Then, the signal captured by the digital receiver is restored through demodulation, carrier synchronization and code synchronization. The restored signal is transmitted to the PC, which is linked with the digital receiver. Hence, we can verify whether the original information is restored from receiver PC. The test setup of the DSSS transceiver is shown in Figure 35, from which we see that the software transceiver is working normally. The signal transmitted form digital transmitter was correctly restored by the digital receiver. The above tests are sufficient to demonstrate the correctness of the designed DSSS software transceiver and the reliability of the proposed code synchronization algorithm.

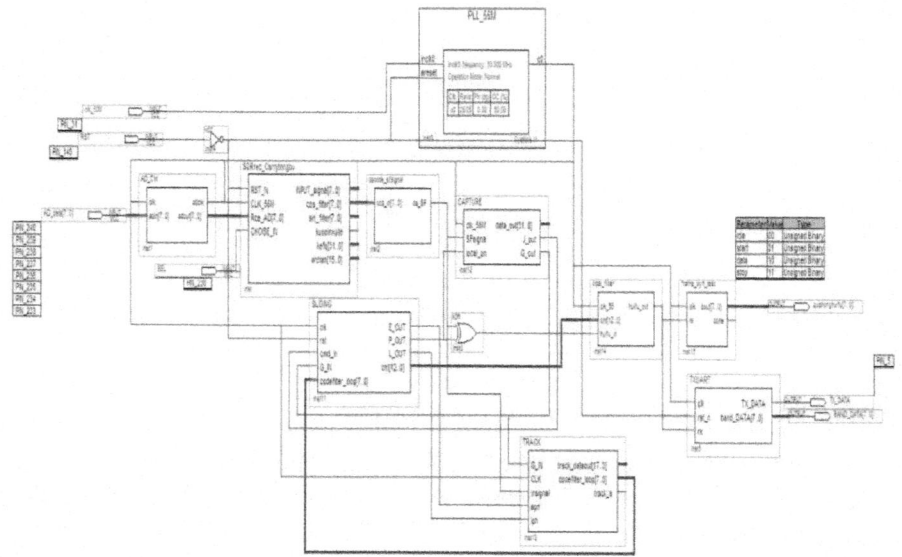

Figure 35: The test setup of the DSSS transceiver.

CONCLUSIONS

In this paper, a serial code acquisition method based on segment correlation is proposed. The proposed code synchronization algorithm can deal with misjudgment caused by the unreasonable data acquisition time effectively. An overall communication system based on the proposed code synchronization algorithm is presented. The M-sequence, BPSK modulation and demodulation, NCO, low-pass filter, carrier synchronization and code synchronization modules were simulated using Matlab. Simulation results show the feasibility of the DSSS transceiver based on the proposed code synchronization algorithm. Each module was then coded and debugged using Verilog. Then, the relevant code was loaded on a FPGA system to verify the overall function of the DSSS transceiver. Finally, a PCB board was designed to verify the overall design.

ACKNOWLEDGMENTS

The work is supported by Qing Lan Project and the National Natural Science Foundation of China under Grant 61572172 and 61401147 and Natural Science Foundation of Jiangsu Province of China, No.BK20131137 and No.BK20140248.

AUTHOR CONTRIBUTIONS

Aohan Li proposed the algorithm and prepared the manuscript. Ziheng Yang was in charge of the overall research. Renji Qi and Feng Zhou were in charge of the implementation of the algorithm. Guangjie Han was in charge of drafted and approved the manuscript.

REFERENCES

1. Zhou, X.; Li, J.; Gou, Y. Software radio technology in spread spectrum communications. In Proceedings of the International Conference on Communication Technology Proceedings, Beijing, China, 21–25 August 2000; pp. 1117–1120.

2. Lin, J.; Wang, H.; Fang, W.; Liu, Y. The research status survey of software radio. *Comput. Meas. Control* 2011, *19*, 2332–2334.

3. Jiang, L. Software Radio System Research of PCI Bus. Master's Thesis, Nanjing University of Aeronautics and Astronautics, Nanjing, China, 2009.

4. Ma, J. Research and Implementation of GMSK Modulation and Demodulation in Software Radio. Master's Thesis, Dalian Maritime University, Dalian, China, 2010.

5. Ma, L. Research on Pseudo-Code Synchronization Technology of DSSS. Master's Thesis, Central South Unicersity, Changsha, China, 2010.

6. Vali, R.; Berber, S.; Nguang, S.K. Accurate derivation of chaos-based acquisition performance in a fading channel.*IEEE Trans. Wirel. Commun.* 2012, *11*, 722–731.

7. Kaddoum, G.; Roviras, D.; Chargé, P.; Fournier-Prunaret, D. Robust synchronization for asynchronous multi-user chaos-based DS-CDMA. *Signal Process.* 2009, *89*, 807–818.

8. Kaddoum, G.; Gagnon, G.; Gagnon, F. Spread spectrum communication system with sequence synchronization unit using chaotic symbolic dynamics modulation. *Int. J. Bifurc. Chaos* 2013, *23*, 365–374.

9.   Li, W.; Li, J.; Chen, Y. Implementation of coding and modulation of an mary orthogonal spread spectrum system. *J. Xidian Univ.* 2000, *27*, 524–527.

10.   He, H.; Yan, L.; Zhao, C. Design of a Spread Spectrum Communication System Based on DSP. In Proceedings of the International Conference on Cyber Technology in Automation Control and Intelligent Systems, Yunnan, China, 20–23 March 2011; pp. 13–26.

11.   Gunnam, K. New optimizations for carrier synchronization in single carrier system. In Proceedings of the IEEE International Conference on Acoustics speech and Signal Processing, Philadelphia, PA, USA, 19–23 March 2005; pp. 661–664.

12.   Stephen, G.W. *Digital Modulation and Coding*; Publishing House of Electronics Industry: Beijing, China, 1998; pp. 197–202.

13.   Bai, Y. Research and FPGA implementation of SDR digital receiver carrier synchronization system. Master's Thesis, Heilongjiang University, Harbin, China, 2012.

14.   Yuan, Q.; Chen, H.; Xu, D. Research on carrier recover algorithm in digital receiver. *J. Harbin Eng. Univ.* 2002, *23*, 57–61.

15.   Feigin, J. Practical costas loop design. *RF Signal Process.* 2002, *6*, 20–36.

16.   Stephen, R. An optimum phase reference detector for fully modulated phase shift keyed signals. *IEEE Trans. Aerosp. Electron. Syst.* 1969, *5*, 627–631.

17.   Tien, M.N. The behavior of costas loop in the presence of space telemetry signals. *IEEE Trans. Commun.* 1992, *40*, 101–107.

18.   Jeong, Y.G.; Kim, J.S.; Kim, M.J. Implementation of code tracking loop for PCS system. In Proceedings of the IEEE Asia Pacific Conference on Circuits and Systems, Kuala Lumpur, Malaysia, 6–9 December 1996; pp. 49–52.

19.   Yang, Y.S.; Lee, S.H.; Yoon, S.S.; Park, H.R. Design and performance analysis of a nonconherent code tracking loop with variable loop bandwidth. In Proceedings of the IEEE International Symposium on Personal Indoor and Mobile Radio Communications, Beijing, China, 7–10 September 2003; pp. 1380–1384.

20.   Huang, X.; Cai, Z.; Hu, S. Design and implementation based on the digital function signal generator of DDS and FPGA. *J. East China Univ. Sci. Technol.* 2009, *32*, 390–393.

21.   Jiang, C.; Tong, Z. A new signal processing modem based on software radio. *J. Electron.* 2003, *14*, 20–24.

22. Mileant, A.; Hinedi, S. Lock detection in costas loops. *IEEE Trans. Commun.* 1992, *40*, 480–483.

23. Lin, Y. A self-normalizing symbol synchronization lock detector for QPSK and BPSK. *IEEE Trans. Wirel. Commun.*2006, *5*, 347–353.

24. Lv, X. Design of Wireless Communication System. Master's Thesis, Changchun University of Science and Technology, Changchun, China, 2004.

25. Zheng, D.; Sun, Y.; Yang, Z. Design and FPGA implementation of SPSP synchronization system based on sliding correlation method. *J. Heilongjiang Univ. Eng.* 2006, *5*, 347–353. (In Chinese).

Chapter 5

NOVEL SYSTEM FOR BITE-FORCE SENSING AND MONITORING BASED ON MAGNETIC NEAR FIELD COMMUNICATION

Andres Diaz Lantada[1], Carlos González Bris[2], Pilar Lafont Morgado[1], and Jesús Sanz Maudes[2]

[1]Mechanical Engineering Department at Universidad Politécnica de Madrid, c/José Gutiérrez Abascal 2, 28006 Madrid, Spain

[2]Conectivity Group, Higher Technical School of Telecommunication Engineering, Universidad Politécnica de Madrid, Av. Complutense 30, 28040 Madrid, Spain

ABSTRACT

Intraoral devices for bite-force sensing have several applications in odontology and maxillofacial surgery, as bite-force measurements provide additional information to help understand the characteristics of bruxism disorders and can also be of help for the evaluation of post-surgical evolution and for comparison of alternative treatments. A new system for measuring human bite forces is proposed in this work. This system has future applications for the monitoring of bruxism events and as a complement for its conventional diagnosis. Bruxism is a pathology consisting of grinding or tight clenching of the upper and lower teeth, which leads to several problems such as lesions to the teeth, headaches, orofacial pain and important disorders of the temporomandibular joint. The prototype uses a magnetic field communication scheme similar to low-frequency radio frequency identification (RFID) technology (NFC). The reader generates a low-frequency magnetic field that is used as the information carrier and powers the sensor. The system is notable because it uses an intra-mouth passive sensor and an external interrogator, which remotely records and processes information regarding a patient's dental activity. This permits a quantitative assessment of bite-force, without requiring intra-mouth batteries, and can provide supplementary information to polysomnographic recordings, current most adequate early diagnostic method, so as to initiate corrective actions before irreversible dental wear appears. In addition to describing the system's operational principles and the manufacture of personalized prototypes, this report will also demonstrate the feasibility of the system and results from the first *in vitro* and *in vivo* trials.

INTRODUCTION

Intraoral devices for bite-force sensing and measurement have several applications for odontology and maxillofacial surgery, as an additional method for assessing the degree and for monitoring of dental and occlusal pathologies and for assessing the functional state of the masticatory system [1]. It is also used for the evaluation of post-surgical evolution and for comparing alternative treatments and their influence on temporomandibular disorders [2]. Because the information derived from such measurements can be used efficiently in clinical tasks, the development of precise and repetitive sensing systems is especially valuable [3]. Several devices (and related patents) have been developed in the past two decades for the aforementioned tasks, allowing intraoral (almost always between occlusal surface of teeth and between upper and lower teeth) bite-force sensing using different mechanical and electronic sensing principles, the most relevant of which are mentioned below.

The use of extensiometric gauges has been reported and validated *in vivo* using animal models, such as primates [4], pigs [5] and dogs [6]. In other cases, several devices have been used to assess the degree of patient satisfaction with dental prostheses, both fixed and removable, in order to validate therapeutic solutions and to propose additional improvements [7–9]. Among the most important studies linked to intraoral bite-force sensing in humans are those related to the study of bruxism activity and temporomandibular joint pathologies [10–12].

Bruxism is a health problem that involves grinding or tightly clenching the upper and lower teeth. Both the grinding and the sliding lead to dental wear and produce a noise during the night that is sufficiently loud to disturb the sleep of anyone nearby, among other symptoms described further on. The phenomenon was introduced in the dental literature as "bruxomania" in 1907, by Pietkiewicz, who described the habit of teeth grinding. The term "bruxism" was introduced by Frohman in 1931 and, in 1936, Miller proposed the use of the term "bruxomania" for daytime grinding and "bruxism" for nighttime grinding. The terms "traumatic neuralgia", "Karolyi effect" and "occlusive habit neurosis" have all been used to refer to some form of teeth grinding or clenching.

The term bruxism is mainly used when the duration and intensity of clenching or grinding activities have a bearing on dental wear and lead to the development of temporomandibular joint (TMJ) problems. However it is important to point out that everyone subconsciously clenches his or her teeth at some time, even healthy population, due to different tooth damages, corporal pains or social conflicts, and this may also be considered bruxism activity [13].

According to studies by the Canadian Sleep Society [14–16], nocturnal bruxism affects 8% of adults and 14% of children. A decrease in the population affected is apparent with age, with around 3% of individuals over 60 being affected. However, for some researchers [17,18] prevalence is around 25%. In terms of gender differences in the affected population, there is no general agreement data.

Main symptoms include lesions to the teeth; problems in the muscles, tissues and other structures surrounding the jaw; headaches; ear pain; reduced neck motility; and, sometimes, irreversible disorders of the jaw joints [13]. All of these symptoms are also collectively described as temporomandibular joint disorders or craniomandibular dysfunction pain syndrome.

Measurements of intraoral bite-force become especially relevant when related to the study of bruxism, especially as a complement for polysomnographic diagnosis, which still is the most adequate procedure [15] for monitoring the evolution of patients and comparing different possible treatments. Special attention should be paid to results from some studies [10–12] that have tried to offer a quantitative definition of bruxism. Results from Nishigawa's study on the bite force produced during bruxism events suggested values as high as 1,100 N, exceeding the maximum voluntary bite force. Pressures on teeth surfaces can reach 40 MPa, sufficient to cause alarming levels of wear and even tooth breakage. In terms of the duration of bruxism events, an average of around 7 seconds has been reported [10,11]. When developing sensors, it is important to distinguish bruxism events from mioclonus or rapid contractions (<0.5 s) of the jaw muscles [11].

One of the main problems associated with the traditional diagnosis of bruxism is that it is frequently made when the teeth are already highly worn and the prognosis of the illness is more severe [19]. Bruxism activity can also be recorded by an electroencephalogram (EEG), as well as by means of electromyography (EMG) and surface electromyography (S-EMG). Video cameras are also used to distinguish between bruxism events of the mioclonus and rapid contractions (<0.5 s) of the jaw muscles [20]. In order to provide additional information for understanding the characteristics of bruxism disorder, as additional information for conventional diagnosis, and to compare different treatment alternatives of bruxism, our research team proposes the use of novel instrumented splints for detecting and recording the intensity and duration of bruxism events.

Explained below are the design, manufacture and trials of a constructed splint, for the monitoring of bruxism activity, which also includes additional telemedicine possibilities, such as patient remote monitoring and control. The reduced-size sensing device allows obtaining an intra-mouth device, which does not cause important alterations of the patient's bite and promotes the quality of final measurements. The use of a wireless external interrogator/reader also allows the collection of the desired information in a more convenient way, when compared to previous proposals from our team [21,22] and to the devices used in other studies [23,24].

DESIGN OF BITE-FORCE SENSING AND MONITORING SYSTEM

Principle of Operation

In this work, we propose a system and explain the development of a prototype that uses a wireless connection between a passive force sensor, located within a splint in the mouth cavity, and an active external unit that energizes the sensor and permanently records all force measurements. This system allows for permanent occlusal force measurements. The prototype uses a magnetic field communication scheme similar to low-frequency radio frequency identification (RFID) technology (NFC). The reader generates a low-frequency magnetic field that is used as the information carrier and powers the sensor.

Unlike other RF signals, the low-frequency magnetic field is not absorbed by biological tissues. This may be an advantage over Bluetooth-based systems [21,22]. However, the communication range is limited to a few centimeters in order to achieve reasonable amplitude values for the excitation signals.

A low-frequency (125 kHz) magnetic carrier field was chosen because it offers good penetration and is not absorbed by biological tissue. Additionally, the inductive coupling of the carrier field allows us to energize passive components and ensures that the transmitting coils and other components will be sufficiently small for intraoral measurement.

Our team is not aware of any rigorous studies of the effects of the low frequency magnetic field in human tissue. Nevertheless, a high intensity low frequency magnetic field is commonly used in passive RFID scanners in consumer retail contactless payment systems, so we initially consider it biologically safe, although losses in metallic fillings need to be considered. However additional long-term testing in animal models should be accomplished before obtaining a product that could be produced on an industrial scale. RF communication suffers from higher absorption in biological tissue, and it does not allow the sensor to be externally powered, thereby requiring the use of a battery. The size of the components is proportional to the operation frequency.

Like RFID, the basic system is composed of an interrogator and a passive sensor. The term passive indicates that the sensor does not use a battery. The interrogator (reader) is usually composed of a microcontroller-based unit and an analogue front-end. The analogue front-end generates a low-frequency (125 kHz) magnetic field that induces a voltage within a tuned LC circuit in the power subsystem of the sensor. This voltage is rectified and used to power the rest of the sensor. This tuned LC circuit is not the same as the LC circuit that is used to communicate with the reader because energy must be permanently supplied to the sensor for it to operate, while the communication module can be independently tuned and detuned.

The small variation produced in the field by the modulation does not affect the energizing module of the sensor. The sensor is only powered when there is sufficient induced voltage in the coil, which implies a maximum distance requirement for the reader. When the sensor is energized, it activates a low power oscillator with an oscillation frequency that is directly related to the resistance value of a force-resistance transducer. A force-dependent oscillator was chosen because it permits continuous monitoring if the interrogator is active and it draws less power in continuous operation than a micro-controlled-based unit that samples, converts and digitizes the readings. In order to transfer the data to the reader, the output of the oscillator is used to bypass a capacitor within an LC tank that is tuned at the frequency of the interrogating field, so that the oscillator tunes and de-tunes the LC circuit at the force dependent oscillating frequency, producing what is commonly termed backscattering or load modulation.

This modulation is sensed at the interrogator as a tiny amplitude modulation (typically around 60 dB), which is recorded using an envelope detector and a locking amplifier. In this manner, the frequency of the detected signal equals the frequency of the oscillating sensor, which depends on the applied force. Frequency modulation was chosen instead of amplitude modulation to make the system less sensitive to relative head movements, which imply distance variations and, correspondingly, undesired amplitude modulation. By contrast, frequency is not affected if the sensor is within communication range.

The oscillator frequency range for the expected transducer values must be selected to match the frequency response of the reader. The demodulated frequency signal is used to evaluate the sensed bite force. The sensor is designed to be placed inside a splint located in the mouth, without remarkably affecting the size of conventional Michigan-type occlusal splints and main types of mouthguards.

Although a patent for the described system [25] with reference number OEPM P200900875 [26] has been granted by the Spanish Patent and Trademark

Office (the University holds the rights), the authors do not intend to develop a commercial product at this time; moreover, significant more funding and extensive *in-vivo* testing using animal models would be mandatory in such a case. A diagram of the components and operating scheme is shown in Figure 1.

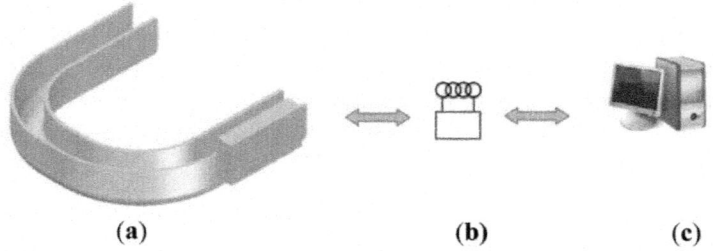

(a)   (b)   (c)

Figure 1: (a) Intraoral splint with passive force sensor. (b) External wireless interrogator. (c) Subsequent data delivery to PC for additional studies.

Intraoral Passive Bite-Force Sensor

The passive sensor, which is the component that must eventually be integrated into the splint, is composed of two different subsystems:

1. The first subsystem is a power subsystem composed of a tuned LC circuit, a Schottky diode rectifier and a low power, low dropout regulator (f.i., MC78LC30) that provides a constant DC voltage with two capacitors at its input and output. This approach avoids using a battery within the mouth. The values of the LC circuit are chosen so that it resonates near 125 kHz.

    Ideally:

$$\sqrt{LC} = 2\pi \, f_0 = 2\pi \, 125 \text{ kHz}$$

Figure 2 depicts the schematic of the energizing subsystem. Subsystem (I) energizes subsystem (II) given sufficient interrogation field intensity.

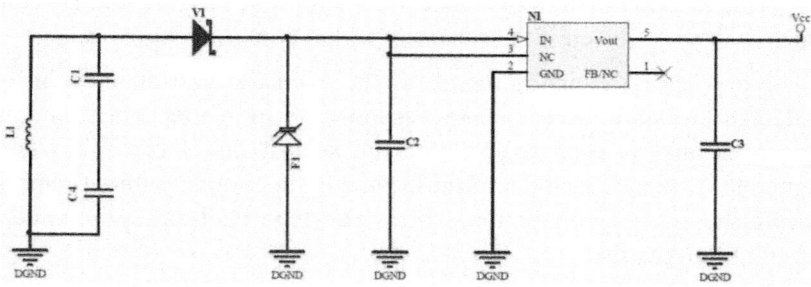

Figure 2: Energizing subsystem of the sensing unit.

2. The second subsystem is a force-sensitive oscillator, which is composed of a low power relaxation oscillator (LMC555 from National Semiconductor) and a few passive components, including a customized for short length (25.4 mm long) commercial force to resistance transducer (ZFLEXA201-100 from Tekscan), which is 0.2 mm thick and features a sensing area of about 0.78 cm^2. Its low thickness is especially appropriate for being integrated in a intraoral device and for not affecting patients' bite performance dramatically. Such thickness is similar to that from other film sensors (normally piezoelectrics) previously used in related research [27].

According to manufacturer (Tekscan) this resistance transducer is constructed using two layers of substrate. This substrate is composed of polyester film (or polyimide in the case of high-temperature applications). On each layer, a conductive material (silver) is applied, followed by a layer of pressure-sensitive ink. Adhesive is then used to laminate the two layers of substrate together to conform the sensor. The silver circle on top of the pressure-sensitive ink defines the "active sensing area". Silver extends from the sensing area to the connectors at the other end of the sensor, forming the conductive leads. The sensor acts as a variable resistor in an electrical circuit. When the sensor is unloaded, its resistance is very high (greater than 5 MΩ). When a force is applied to the sensor, the contact between conductive particles of the pressure-sensitive ink is promoted and the resistance decreases. Additional information on the working principle and conditioning proposals can be obtained from specification sheets available at Tekscan's website (www.tekscan.com, last access 1 August 2012).

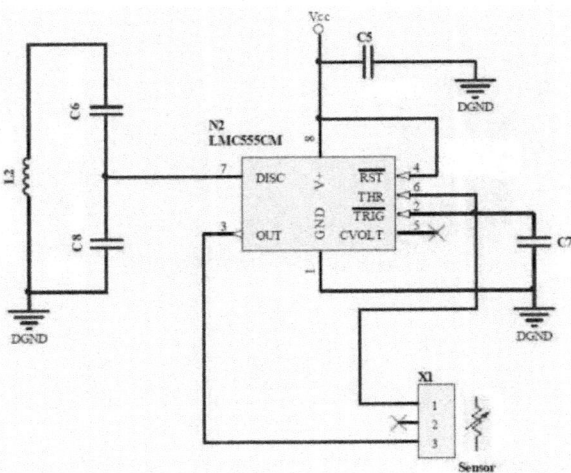

Figure 3: Force-sensitive oscillator subsystem of the sensing unit.

Such a transducer must be placed in the splint, where it converts the occlusal force into a resistance, which determines the oscillation frequency. When force is applied to the transducer, its resistance decreases and such decrease produces a frequency shift in the oscillator. The output of the relaxation oscillator drives a tuned LC circuit, which is tuned and de-tuned at the oscillation frequency. The sensor oscillation frequency range of this component is chosen to match the response of the filter located at the reader element, and the maximum oscillation frequency is selected to match the minimum resistance of the force transducer, which presumably happens when a bruxism event is occurring. When operating, this component consumes about 250 μW. Figure 3 shows the schematic of the force-sensitive oscillator.

Figure 4 (upper panel) shows two prototypes of the passive force sensor. The implemented sensing unit made up of a printed circuit that is 2 cm long, 5 mm wide and 5 mm thick and a force to resistance transducer, which is trimmed to a length of about 25.4 mm, which is soldered to the printed circuit so that it appears as a protruding tongue. This unit is integrated in the splint.

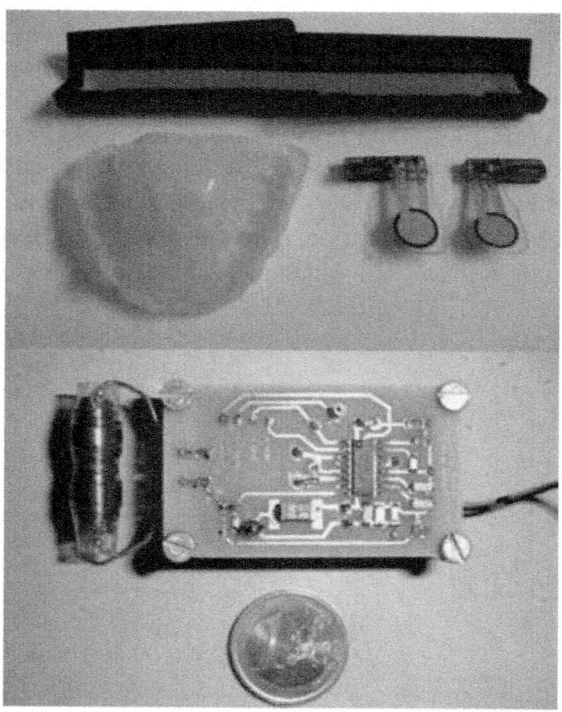

Figure 4: Splint, passive sensors with transducers and photopolymerizable resin for encapsulation (upper image). View of the external interrogator prototype (lower image).

External Interrogator

A reader like the one used in this research is typically composed of a microcontroller and an analogue front-end unit. This unit generates the field that energizes the passive sensor and detects the modulated signal that is produced at the excitation coil by the tuning and de-tuning of the sensor tank circuit. In our first implementation, we have only used a front-end unit that, once powered, permanently generates the magnetic field using a series LC circuit. The microcontroller permits control of the excitation activity duty cycle of the front-end unit and readout of the demodulated signal. The demodulated output provided by the front-end unit was monitored using an oscilloscope in our case. In the prototype, we used an (EM4095) analogue front-end with external passive components.

The components were selected to generate a 60 Vpp at the excitation coil. An antenna current of 230 mArms corresponds to this value, which is below the 300 mArms limit listed in the specifications. A supply current of 175 mA was measured under the operating conditions with a voltage of 5 V. This value results in a power dissipation of 875 mW, which corresponds to about 200 mW in the chip and 700 mW in the external current limiting resistor located in the excitation tank. The chip temperature increase is about 15 °C, which is less than the specified maximum temperature value (100 °C) at normal room temperature. A set of four standard rechargeable AA batteries lasts about 12 hours if the unit is permanently in the 'on' state. The values of certain components within the circuit determine the filtering behavior of the demodulator. All components were selected to maximize sensitivity at a frequency of nearly 2 kHz, which corresponds to the force threshold applied to the transducer during bruxism events. Figure 4 (lower panel) shows the interrogator/reader prototype used to evaluate the system.

Prelimminary in Vitro Calibration

Prototype passive sensor behavior was characterized using a workbench based on pneumatic muscles (Festo MAS-20-160N-AA-MC-K) capable of applying bite forces in the range of 0–1,400 N. The measurement range was selected because the occlusal force during a bruxism event is expected to reach around 1,000–1,200 N, in accordance with the values that have been measured during events in controlled environments [22,27]. The final value will depend on each specific patient.

To obtain the desired calibration curve, the transducer of the sensor was placed on the workbench between two metallic pieces. Force was applied using the pneumatic muscle, while the sensor was activated by the interrogator and each force level was repeated three times. The output demodulated signal provided by the reader was measured using the screen trace of an Agilent Infiniium 54832D oscilloscope, together with a USB data acquisition board, Measurement Computing model LS 1208.

Applied force can be calculated as a function of the supply pressure applied by taking into account the geometry of the workbench and the response characteristics of the muscle. Figure 5 illustrates the results and for each force level studied a standard deviation below 5% was measured among repetitions.

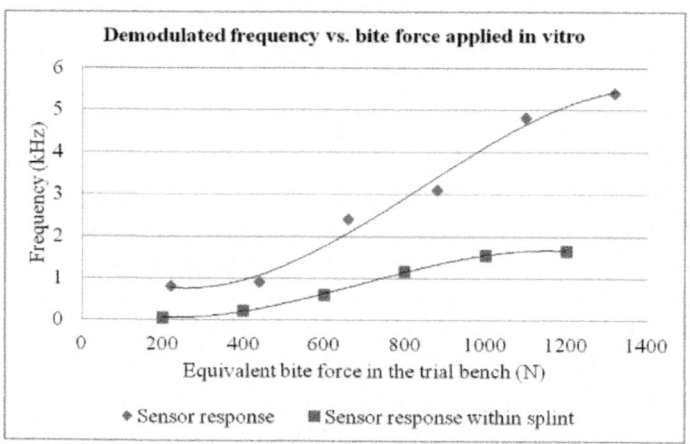

Figure 5: Sensor calibration curve before encapsulation and once encapsulated within the splint.

The calibration curve of the sensor, before encapsulation within the intraoral protection splint, proves to be adequately linear in the measurement range of typical bruxism events (350–1,000 N). The observed communication range is about 5 cm if the sensor is energized using a constant power supply (*i.e.*, using an external supply or battery). Nevertheless, the limited communication range can be considered an advantage because it avoids interference with other devices.

It is important to note that final sensor encapsulation within the splint changes measurement, because the photopolimerizable resin used (see Section 3) absobs the applied force, so a final calibration step, once the sensor is included within the splint, is needed for finally relating measured frequency shift and actual bite force. For such final calibration an *ad hoc* workbench can be employed, with pneumatic muscles acting on personalized plaster models reproducing the teeth of the simulated patient, which finally act on the splint, as has already used for our previous developments [22,27]. The effect of sensor encapsulation within the splint can hence be assessed and is represented below in Figure 5 as final device characterization step.

PROTOTYPE MANUFACTURE

The idea of including the passive force sensors within an intraoral splint responds to a need to locate the transducer in a fixed location between the teeth of the patient, so as to obtain repeatable measurements. At the same time, such polymeric splints help to protect and isolate the system. The manufacturing process is explained below.

First, two silicone moulds of the upper and lower teeth of the patient are obtained by rapid form copying. Plaster models of a patient's teeth are subsequently obtained by casting with the help of the aforementioned silicone moulds. The base layer of the splint is obtained by vacuum/pressure thermoforming against the plaster models using (Erkodur) thermoplastic, 1.5 mm-wide polymeric discs. This base layer gives structural support to the whole system and after careful polishing, can be secured to the upper teeth of the patient. The personalized design provides additional comfort. The electronic components are encapsulated and adhered to the base layer of the splint with the help of a Kuss Dental Kit.

Such splint manufacturing kits include the photopolymerizable resin "Delta Splint", which can be conformed manually to the desired forms, glued to the base layer with the help of "Delta Bond" and finally cured by UV exposure. Using the "Delta Splint" resin, the passive force sensor is first encapsulated. After curing it is bonded to the base layer so that it can be positioned in the vestibule of the mouth. Additional protective layers of photopolymerizable material can be included to enhance results. The step-by-step manufacturing process is shown in Figures 6 and 7.

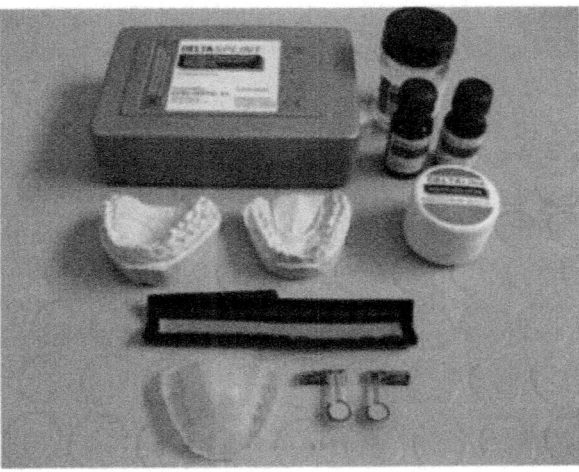

Figure 6: Components for manufacturing the intraoral splint.

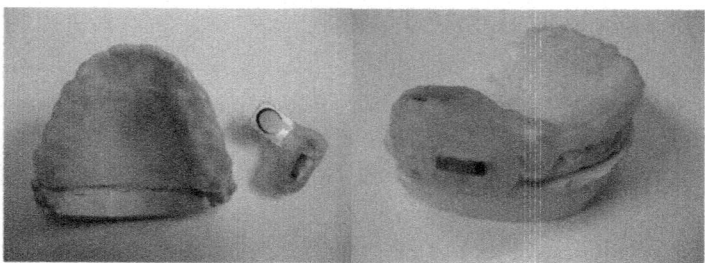

Figure 7: Splint prototype manufacturing process: Instrumentation of the intra-oral splint.

It is important to note that the use of the mentioned photopolymerizable resin allow encapsulating at room temperature, what proves to be very positive for protecting the electronics from non-desired damages that usually appear when using high-temperature processing stages. We have not appreciated any effect due to the resin shrinkage during curing, as the layer of resin applied is very thin (around 1 mm), although further specific assessment of such influence may be of interest if the system is encapsulated in other biodevices requiring for instance spherical or cylindrical encapsulations.

The exposed process is also noteworthy due to its versatility, helping to find the most ergonomic location for the intra-oral components, and is especially well-suited for research studies related to pre-commercial devices, in which final distribution of components is not yet defined. Regarding industrialized manufacturing processes previous studies from our team have also analyzed the use of semi-automatic thermoforming between several layers of acrylic thermoplastic polymeric discs for encapsulation [22], valid for the production of short and medium series, although not so versatile for prototype manufacture and preliminary validation stages.

Final device size is similar to that of recent devices for intra oral monitoring [23,24] and the *in vivo* trials, explained further on, help to validate our approach in a preliminary way, before carrying out a more exhaustive clinical study for analyzing additional factors of influence such as, sex, age, professional activity, combined pathologies, among others, also with the patients during their sleep. To our knowledge it is the first intraoral wireless and battery-less intraoral devices with these diagnostic capabilities, what helps also to enhance usability and final device safety.

IN VIVO TRIALS FOR VALIDATION

After the bite-force sensing system prototype was manufactured, *in vivo* trials were carried out to validate its capabilities and advantages compared with existing devices, as well as to evaluate principal required improvements. Figure 8 shows the *in vivo* validation, including the intraoral constructed splint from Figure 7 (located inside researcher's mouth, so it cannot be seen) and the external interrogator from Figure 4, place on the researcher's face and connected by cable to an Agilent Infiniium 54832D oscilloscope as the frequency signal detector, in which the frequency of the signal was read from the screen output.

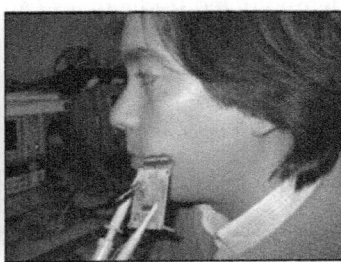

Figure 8: *In vivo* validation trials.

The external interrogator was always placed against the face of the patient (30 year old researcher without occlusal problems) so that the relative position between the passive sensor and the interrogator would not change. It is important to note again that no wired connection was used between the intraoral instrumented splint and the external interrogator, as the system works by wireless magnetic field communication. The cables seen in Figure 8 just connect the external interrogator to the battery package and to the oscilloscope for measurement recording.

The system proved to be reliable and accurate with a communication range of around 3 cm, even though the sensor and reader operated through human flesh. A higher range can be achieved using a higher excitation. This range cannot be easily obtained using other wireless approaches, such as Bluetooth/ ZigBee. The main results are summarized in Figure 9 and explained further on.

During the *in vivo* trials, bruxism events were deliberately simulated to verify the capability of the system to quantitatively assess events of different magnitudes. The researcher was first asked to consciously bite with maximum force. Afterwards, he was requested to apply bite forces of different magnitudes relative to the initial bite force applied (75%, 50%, 25% and 10%) to assess system behavior within the mouth of a person and its quantification ability, as shown in Figure 9. Such *in vivo* trial was intended to provide a conscious simulation of different bruxism events, as well as to assess the mechanical stability and performance of the prototype inside patient's mouth. Each bite force level was repeated five times and maintained for 5 seconds, with a rest interval of 10 seconds between each measurement, so as to verify repeatability of the measuring system. For each bite force level, final standard deviation obtained was below 15%, what we consider adequate for prototype quality validation and taking into account the limitations of *in vivo* probing, highly useful for assessing the device integrity and response in real working conditions, but somehow limited when trying to simulated similar events several times. With the calibration curve of the constructed splint (see Figure 5), such demodulated frequencies can be related to actual bite forces.

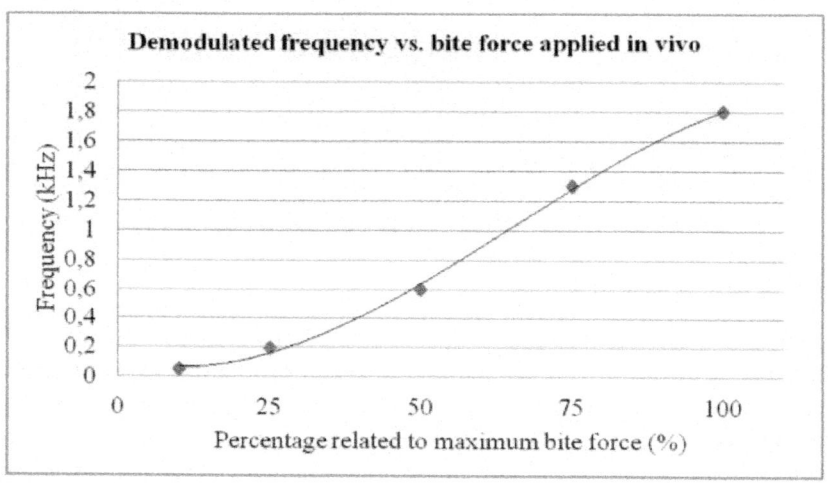

Figure 9: Results summary. Frequency *vs.* applied force (related to maximum bite force).

In addition the transversal component of force performed during bruxism events was also simulated, but the system did not response in a predictable way, as the transducer used here is designed for the measurement compressive stresses. A measurement of both vertical and horizontal forces can be obtained by using a different kind of transducer, such as piezoelectric polymers (*i.e.*, PVDF, due to its non-zero value of d_{33}, d_{31} and d_{32} piezoelectric coefficients), as our previous studies have shown [27]. However the use of polymeric piezoelectric sensors requires more complex conditioning, including a charge amplifier, and signal is less robust to the influence of movements and external signals. In addition the piezoelectricity of these polymers is consequence of their being ferroelectric and, therefore also pyroelectric, so the final signal is also highly influenced by temperature changes [28].

Future trials should address a possible combination of different kinds of sensors for improving the diagnostic capabilities of the system, specially focusing on providing additional information about horizontal forces (interesting for transversal bruxism) and further device miniaturization should be pursued to allow more comfortable recording of bite force of bruxism patients during sleeping. Our advice would be to change the current external reader morphology, using vertical axis flat excitation coil, which could be located like a patch on patient's face. Integration of components would significantly reduce external reader's size and additionally improve the comfort of all-night monitoring.

We also noted that, due to the additional protection of the transducer provided by the splint, the values of the demodulated frequencies measured *in vivo* are smaller than those obtained with the initial *in vitro* calibration using the trial bench. As instrumented splints can be manufactured using different polymeric materials, with different elasticity, the behavior of the system after manufacture should always be calibrated *in vitro*, using for instance a pneumatic workbench [22,27], in accordance with the desired personalized approach.

In any case, linearity of the system is especially favorable in the conventional bruxism event range (from 30% to 90% of maximum bite force, *i.e.*, around 300 to 1,300 N). The results of our trials with a personalized splint show that it is possible to detect bruxism events of different intensities and durations, which, combined with the ability to record and store the data, converts the system into a "Holter" for bruxism diagnosis/monitoring and into an additional tool for evaluating clenching forces or bruxism activity.

CONCLUSIONS

This paper has focused on the complete development of a splint to assess intraoral clenching forces, as an aid or additional resource for diagnosing bruxism and other occlusal pathologies. The system includes an active interrogator/reader and a passive sensor that can be used to detect and record bruxism events. The system uses a low-frequency magnetic field to energise the passive sensor, located within the splint, and to detect information about the force magnitude before transmitting it via a communications protocol that is similar to RFID. The passive sensor extracts the energy that it needs to operate from the excitatory field and converts the sensed force into a frequency signal that is sent wirelessly to the reader using load modulation. Additional remote monitoring activities can be directly implemented *i.e.*, via mobile phone.

A prototype of the passive sensor/active reader has been built to evaluate and characterize the behaviour of the sensing system. Our results indicate that the sensor can be used to continuously monitor force as a function of time, although it can also be enabled/disabled periodically by a controller. The prototype is only intended to evaluate the viability of the approach; final integrated devices may be the size of a tooth. The system, to our knowledge, is unique in its use of magnetic fields to energise the passive force sensor, which allows for permanent function and eliminates the requirement for intraoral batteries. Hence, the proposed system is much safer than previous instrumented splints with communication capabilities that demand an intraoral energy supply.

Additionally, we highlight the possibility of obtaining a device, whose measurements will provide additional information for diagnosis of bruxism and conservative treatments can be taken before the appearance of irreversible dental wear.

The first *in vitro* and *in vivo* trials to date have served to demonstrate the feasibility of the system in accordance with our proposed objectives. Finally, it should be emphasized that a similar working principle can be applied to the development of several biomedical and biomechanical devices that can benefit from force measuring capabilities and wireless communication, even working through patient's flesh or tissues, such as instrumented intelligent prosthesis or health monitoring systems.

Future efforts will be directed to a more systematic *in vivo* study, in order to assess the actual clinical effectiveness of the proposed device, aiming at an earlier diagnosis of craniofacial and occlusal disorders, including bruxism.

ACKNOWLEDGMENTS

Authors would like to acknowledge the support of the Spanish Ministry of Science and Education through the following research project: "FEMAB: Férula microinstrumentada anti-bruxista" Project (with Reference Number CIT-020400-2005-17). We also acknowledge the reviewers for their positive comments and proposals for improvement, which have helped to provide a more detailed and clear explanation, as well as to propose future directions. Finally, the authors are not aware of any potential conflicts of interest.

REFERENCES

1. Koc, D.; Dogan, A.; Bek, B. Bite force and influential factors on bite measurement: A literature review. *Eur. J. Dent.* 2010, *4*, 223–232.

2. Pereira, L.J.; Pastore, M.G.; Bonjardim, L.R.; Castelo, P.M.; Gavião, M.B.D. Molar bite force and its correlation with signs of temporomandibular dysfunction in mixed and permanent dentition. *J. Oral Rehabil.* 2007, *34*, 759–766.

3. Pereira-Cenci, T.; Pereira, L.J.; Cenci, M.S.; Bonachela, W.C.; Del Bel Cury, A.A. Maximal bite force and its association with temporomandibular disorders. *Braz. Dent. J.* 2007, *18*, 65–68.

4. Dechow, P.; Carlson, D.S. A method of bite force measurement in primates. *J. Biomech.* 2003, *16*, 797–802.

5. Bousdras, V.; Cunningham, J.L.; Ferguson-Pell, M.; Bamber, M.A.; Sindet-Pedersen, S.; Blunn, G.; Goodship, A.E. A novel approach to bite force measurements in a porcine model *in vivo*. *Int. J. Oral Maxillofac. Surg.* 2006, *35*, 663–667.

6. Lindner, D.L.; Marretta, S.M.; Pijanowski, G.J.; Johnson, A.L.; Smith, C.W. Measurement of bite force in dogs: A pilot study. *J. Vet. Dent.* 1995, *12*, 49–52.

7. Rismanchian, M.; Mostajeran, E. Evaluation of maximum bite force and satisfaction in patients with conventional full denture and over denture supported by mandibular dental implant. *J. Isfahan Dent. Sch.* 2007, *2*, 1358–1358.

8. Tortopidis, D.; Lyons, M.F.; Baxendale, R.H.; Gilmour, W.H. The variability of bite force measurement between sessions, in different positions within the dental arch. *J. Oral Rehabil.* 1998, *25*, 681–686.

9. Fernandes, C.P.; Glantz, P.; Svensson, S.A.; Bergmark, A. A novel sensor for bite force determinations. *Dent. Mater.*2003, *19*, 118–126.

10. Nishigawa, K.; Bando, E.; Nakano, M. Quantitative study of bite force during sleep associated bruxism. *J. Oral Rehabil.*2001, *28*, 485–491.

11. Baba, K. Bruxism force detection by a piezoelectric film-based recording device in sleeping humans. *J. Orofac. Pain*2003, *17*, 58–64.

12. Cosme, D.; Baldisserotto, S.M.; Canabarro, S.; Shinkai, R.S. Bruxism and voluntary maximal bite force in young dentate adults. *Int. J. Prosthodont.* 2005, *18*, 328–332.

13. Lavigne, G.; Khoury, S.; Abe, S.; Yamaguchi, T.; Raphael, K. Bruxism physiology and pathology: An overview for clinicians. *J. Oral Rehabil.* 2008, *35*, 476–494.

14. Lavigne, G.; Montplaisir, J.V. Bruxism: Epidemiology, diagnosis, pathophysiology and pharmacology. *Adv. Pain Res. Treat.* 1995, *21*, 387–404.

15. Lavigne, G.; Dao, T.T. Sleep bruxism: Validity of clinical research diagnostic criteria in a controlled polysomnographic study. *J. Dent. Res.* 1995, *75*, 546–552.

16. Lavigne, G.; Kato, T.; Kolta, A.; Sessle, B.J. Neurobiological mechanisms involved in sleep bruxism. *Crit. Rev. Oral Biol. Med.* 2003, *14*, 30–46.

17. Melis, M.; Abou-Atme, Y.S. Prevalence of bruxism awareness in a Sardinian population. *J. Craniomandib. Pract.* 2003,*21*, 144–151.

18. Abou-Atme, Y.S.; Melis, M.; Zawawi, K.H. Bruxism prevalence in a selective Lebanese population. *J. Leban. Dent. Assoc.* 2004, *41*, 31–35.

19. Scrivani, S.J.; Keith, D.A.; Kaban, L.B. Temporomandibular disorders, review article. *N. Engl. J. Med.* 2008, *359*, 2693–2705.

20. Baba, K.; Tsukiyama, Y.; Yamazaki, M.; Clark, G.T. A review of temporomandibular disorder diagnostic techniques. *J. Prosthet. Dent.* 2001, *86*, 184–194.

21. Lafont Morgado, P.; Díaz Lantada, A. System for the Diagnosis and Detection of Bruxism and Other Occlusal Pathologies. PCT/ES2008/000498, 15 July 2008.

22. Lafont Morgado, P.; Díaz Lantada, A. Instrumented Splint for the Diagnosis of Bruxism. Proceedings of the Biodevices First International Conference on Biomedical Electronics and Devices. IEEE Engineering in Medicine and Biology Society, Madeira, Portugal, 28–31 January 2008.

23. Clauss, J.; Wolf-Dieter, S.; Wolf, B. *In-vivo* Monitoring of Bruxism with an Intelligent Tooth Splint—Reliability and Validity. Proceedings of the World Congress on Medical Physics and Biomedical Engineering, Munich, Germany, 7–12 September 2009. [CrossRef]

24. Kim, J.H.; Mc Auliffe, P.; O'Connell, B.; Diamond, D.; Lau, K.T. Development of wireless bruxism monitoring device based on pressure-sensitive polymer composites. *Sens. Actuators A Phys.* 2010, *163*, 486–492.

25. González Bris, C.; Díaz Lantada, A. A Wearable Passive Force Sensor/Active Interrogator intended for Intra-splint Use for the Detection and Recording of Bruxism. Proceedings of the 3rd International Conference on Pervasive Computing Technologies for Healthcare, London, UK, 1–3 April 2009.

26. González Bris, C.; Lafont Morgado, P. Sistema de telemetría mediante comunicación inalámbrica empleando campo magnético sensor pasivo interrogador activo para diagnóstico y detección de episodios bruxistas. OEPM P200900875, 31 March 2009.

27. Díaz Lantada, A. Metodología Para el Desarrollo de Dispositivos Médicos Basados en el Empleo de Polímeros Activos Como Sensores y Actuadores. Ph.D. Dissertation, Universidad Politécnica de Madrid, Madrid, Spain, 2009.

28. Díaz Lantada, A.; Lafont Morgado, P.; Del Olmo, H.H.; Lorenzo Yustos, H.; Echavarri Otero, J.; Munoz-Guijosa, J.M.; Muñoz-García, J.; Muñoz Sanz, J.L. Modelling and Trials of Pyroelectric Sensors for Improving Its Application for Biodevices. Proceedings of the Biodevices Second International Conference on Biomedical Electronics and Devices. IEEE Engineering in Medicine and Biology Society, Porto, Portugal, 14–17 January 2009.

Chapter 6

A FRAMEWORK FOR UWB-BASED COMMUNICATION AND LOCATION TRACKING SYSTEMS FOR WIRELESS SENSOR NETWORKS

Juan Chóliz, Ángela Hernández and Antonio Valdovinos

Research Institute of Engineering in Aragón, I3A, University of Zaragoza, C/María de Luna 3, Zaragoza 50018, Spain

ABSTRACT

Ultra wideband (UWB) radio technology is nowadays one of the most promising technologies for medium-short range communications. It has a wide range of applications including Wireless Sensor Networks (WSN) with simultaneous data transmission and location tracking. The combination of location and data transmission is important in order to increase flexibility and reduce the cost and complexity of the system deployment. In this scenario, accuracy is not the only evaluation criteria, but also the amount of resources associated to the location service, as it has an impact not only on the location capacity of the system but also on the sensor data transmission capacity. Although several studies can be found in the literature addressing UWB-based localization, these studies mainly focus on distance estimation and position calculation algorithms. Practical aspects such as the design of the functional architecture, the procedure for the transmission of the associated information between the different elements of the system, and the need of tracking multiple terminals simultaneously in various application scenarios, are generally omitted. This paper provides a complete system level evaluation of a UWB-based communication and location system for Wireless Sensor Networks, including aspects such as UWB-based ranging, tracking algorithms, latency, target mobility and MAC layer design. With this purpose, a custom simulator has been developed, and results with real UWB equipment are presented too.

INTRODUCTION

The use of location and tracking information is an excellent tool to improve productivity and to optimize resource management in a wide range of sectors: industrial, medical, home-automation or military. Whereas satellite systems, *i.e.*, GPS, are widely used in outdoor applications such as vehicle navigation, fleet management or emergency call localization, there are multiple alternatives for the development of indoor Location & Tracking (LT) systems. Despite not being specifically designed for that purpose, various widespread radio technologies such as cellular (GSM, UMTS, LTE) and short-medium range wireless systems (WiFi, Bluetooth, ZigBee, RFID), may provide location information with different levels of accuracy, range and complexity [1,2]. Within this group of short range radio systems, Ultra-Wideband (UWB) stands out providing high accuracy on distance estimation with remarkable features concerning size and power consumption and allowing simultaneous location and data transmission [3].

IR (Impulse Radio) UWB communication systems are based on the transmission of very short duration pulses, which originate very high bandwidth signals. The short duration of the pulses allows a high level of accuracy in time of arrival estimation and as a result a centimeter-level resolution in distance estimation (ranging). Furthermore, due to the short duration of the transmitted pulses, UWB provides unmatched performance on multipath and NLOS environments [4]. In addition, low complexity and low power consumption of UWB transceivers is essential in order to design battery-powered sensors.

In general, location determination comprises two phases, angle and/or distance estimation and position calculation. Angles and distances between the element to be located and some fixed reference nodes can be estimated based on the measurement of different parameters such as Angle of Arrival (AOA), Received Signal Strength Indication (RSSI) and Time of Arrival (TOA) of reference signals exchanged between them. In particular, TOA estimation requires the exchange of ranging frames between the element to be located and the reference nodes, which entails that some temporal resources must be dedicated to location and that a non-negligible latency is associated to the position update process. On the other hand, several algorithms can be used to compute the position according to the estimated distances or angles, including geometry-based (triangulation, trilateration, multidimensional scaling), least square and cost function minimization, fingerprint and Bayesian techniques (Kalman and particle filters).

Extensive research has focused on the design of distance estimation and position calculation algorithms in the last few years [5–8]. Nevertheless, in general these studies focus on algorithm optimization, and simple scenarios with a single terminal and a few previously defined reference nodes are considered. Only a few studies address practical aspects such as the design of the functional architecture, the procedure for the transmission of the associated information between the different elements of the system, and the need of tracking multiple terminals simultaneously in various application scenarios. These aspects would lead to consider the amount of resources associated to the global location service as a quality parameter. Moreover, a rigorous algorithm evaluation requires using a dynamic scenario with multiple mobile terminals and reference nodes.

On the other hand, although a few UWB-based LT systems can already be found in the market, these systems are proprietary solutions and only use UWB for distance estimation, while communication between the different elements involved in the positioning system is done using other technologies, generally wired. In contrast, UWB systems combining location and data transmission (for example in the framework of IEEE 802.15.4a standard) would increase flexibility reducing the cost and complexity of the system deployment.

IEEE 802.15.4 standard offers the fundamental lower network layers of wireless personal area networks (WPAN) focusing on low-cost, low-speed ubiquitous communication between devices with little to no underlying infrastructure [9]. It provides support for *ad hoc* networks capable of performing self-management and organization, aimed at Wireless Sensor Network (WSN) applications. IEEE 802.15.4a-2007 enhanced the standard specifying two additional PHYs: Chirp Spread Spectrum and Direct Sequence UWB, which enhances the standard with the accurate distance estimation capability of UWB [10]. This way, a UWB-based Wireless Sensor Network (e.g., a fire detection sensor network) could be simultaneously used to provide mobile users tracking. In this scenario, it becomes clear that accuracy is not the only evaluation criteria of LT systems, but also the amount of resources associated to the location service, as it has an impact not only on the location capacity of the system, but also on the consequent reduction of the available data rate for sensor data communication.

The combination of wireless sensor networks with UWB accurate location capabilities enables a wide variety of application scenarios. For example, in order to guarantee the safety of workers in dangerous environments (electrical substations, fires, accidents, *etc.*) tracking their position and monitoring at the same time the level of different parameters (electric field, carbon monoxide, radiation, *etc.*). Another example is sports tracking, in order to provide a complete monitoring of performance (distance travelled, average and peak speed, acceleration, *etc.*) and biometric information (heart rate, blood pressure, *etc.*). In industrial applications, the location of a certain product in the assembly line or in a warehouse could be monitored, together with its temperature, humidity, *etc.*

The main objective of this paper is to design a communication and location tracking system for Wireless Sensor Networks based on Ultra-Wideband technology and to provide a complete system-level evaluation. Besides distance estimation and location and tracking algorithms, system level evaluation includes aspects such as target mobility, functional architecture, distribution of the information related to the location function and latency of the position update process. These aspects are not usually considered in the existing studies, but have a great impact on the capacity of systems combining data transmission and location.

The paper is structured as follows. Section 2 summarizes the literature related to the presented study. In Section 3 the proposed communication and location tracking UWB system is described, including the network topology, the tracking functional architecture, the distribution and acquisition of location information and the tracking function implementation. In Section 4 the system performance is evaluated using a custom self-developed simulator, and the different system design alternatives are assessed. Measurements with real UWB equipment are also provided. Section 5 summarizes the main conclusions.

RELATED WORK

As it was previously mentioned, most of the previous works related to UWB location and tracking systems focus on distance estimation techniques and location and tracking algorithms, but only a few of them address practical aspects such as the transmission of the information associated to location among the different elements of the system. Some proposals of UWB-based communication and location tracking systems can be found in the literature. In [11] a UWB-based system for indoor location services is introduced, which relies on a three-tier hierarchical sensor architecture to cover a large indoor space, also defining the communication between the different tiers and the location procedure. In [12] the design of a UWB-based *ad hoc* network for search and rescue operations in disaster zones is presented, defining the network architecture, physical entities and a complete protocol stack, from the physical layer up to the application layer. In [13] a group of communication protocols and localization algorithms for wireless sensor networks in coal mine environments is proposed, specifically a new UWB coding method, an ALOHA-type channel access method and a message exchange protocol to collect location information. Finally, in [14] an overview of an IR UWB open prototyping platform that illustrates a fully integrated solution from physical layer up to application layer is provided. However, these works [11–14] mainly focus on the system description, including in some cases a very basic system evaluation. Based on the system proposed in [14], this paper presents a thorough analysis and evaluation of different aspects such as the functional architecture, the acquisition and distribution of the information associated to location and the effect of position update latency and target mobility.

The optimization of MAC layer design for simultaneous location and communication has also been addressed in different works. In [15] an early approach to UWB MAC layer issues for location and tracking applications is provided and various UWB system architectures, MAC schemes and network solutions are discussed, although no evaluation of the proposals is provided. In [16] a MAC layer design for IR UWB location networks is proposed in order to determine the locations of a network of stationary reference nodes and mobile nodes deployed in an *ad hoc* manner based on a small number of fixed anchors with known locations. Location and range information is propagated through the network of reference nodes to periodically estimate the locations of the mobile nodes, which entails a convergence time and associated throughput. On the contrary, the tracking system proposed in the present work considers that the locations of the reference nodes are known, whether they are known a priori or determined in a prior set-up phase. This way the convergence time and associated throughput are minimized, thus providing fast tracking of the mobile nodes. The MAC layer design we have considered as the basis for the proposed UWB communication and location tracking system was presented in [17]. This MAC is based on the IEEE 802.15.4 standard [9], although it deviates from the standard in a few areas such as the support to peer-to-peer communications, the usage of guaranteed time slots for data transmissions and dedicated time slots for ranging and allocation requests, the definition of a relaying functionality and the specification of ranging and localization procedures. In [18] the performance of this MAC is studied under the point of view of tracking, evaluating the time delay necessary to collect the ranging information as a function of the number of mobiles in the network. Furthermore, a few enhancements are proposed in order to minimize the exchange of packets necessary to update the ranging information. However, this study assumes the existence of physical connectivity between all the nodes and does not take into account the resource constraints of the MAC layer design, aspects that are considered in this study, which also evaluates the performance degradation in terms of accuracy due to the latency associated to the position update process.

WIRELESS SENSOR COMMUNICATION AND LOCATION TRACKING SYSTEM PROPOSAL

The proposed communication and location tracking UWB system aims to enable wireless data communication within a network of sensors and at the same time to track walking users in wide indoor areas with accuracy below 1 m. The system is composed of multiple UWB picocells. Each picocell is composed of mobile nodes to be tracked (targets) and fixed sensor nodes with known positions (anchors). Distances between the target and the anchor nodes are estimated through a ranging frame exchange. Estimated distances are sent to location controllers (LC), which are the functional units that execute the tracking algorithm to obtain the estimated position of the targets. The main characteristics of the network and the different options considered are detailed below.

Network Topology and PHY/MAC Structure

The application scenario is covered by multiple UWB picocells, although for simplicity a single picocell is considered. PHY and MAC layers of an open IR-UWB platform described in [14] are assumed. This platform is based on the 802.15.4a standard [10], although it is not fully compliant. Table 1 summarizes the main PHY and MAC parameters considered.

Table 1: PHY and MAC parameters

Parameter	Value
Frequency range	3.5–4.5 GHz
Symbol duration	2.88 μs
Raw bit rate	347 kbps
Slot length	160 bytes
Slot duration	3.686 ms
Maximum superframe length	53 slots
Beacon Interval	195.379 ms
Maximum Beacon Period length	12 slots
Max. Topology Management Period length	3 slots (12 subslots)
Maximum CFP length	26 slots
Number of slots for data communication	8 slots
Number of slots for ranging	12 slots
GTS request period length	6 slots (12 subslots)
Maximum CAP length	12 slots (48 subslots)

The picocell topology is mesh centralized, as shown in Figure 1. A picocell coordinator transmits beacon frames for common superframe synchronization and handles the scheduling procedures. Then, a scheduling tree is built and used to transport beacon and command frames, which are relayed from the picocell coordinator to any node in the picocell. Finally, it becomes a meshed scheduling tree by enabling the transmission out of the tree for the data, ranging and hello frames.

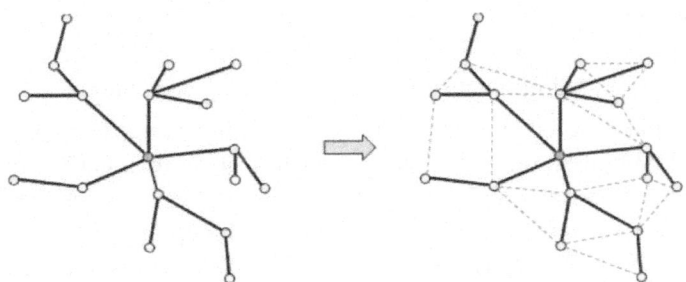

Figure 1: Mesh centralized topology.

The MAC superframe is divided into timeslots that are grouped into different periods, as it is shown in Figure 2.

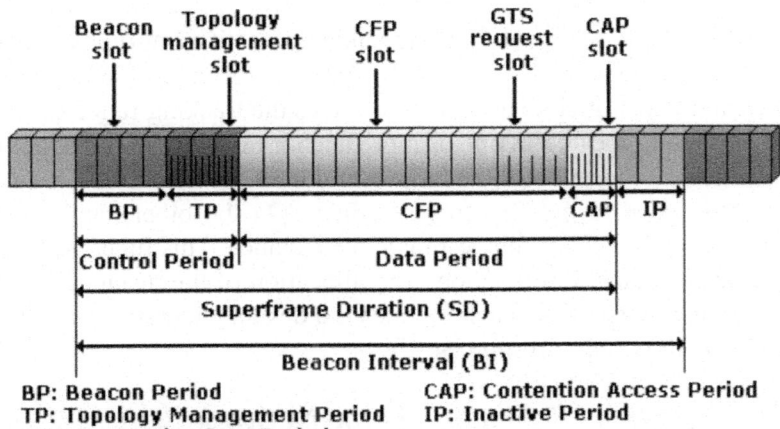

Figure 2: Proposed MAC superframe structure.

- Beacon period: Used for the beacon alignment. The first beacon slot is reserved for the coordinator.

- Topology Management Period: Used for the periodic broadcast of hello frames from each node. This way the neighborhood is known locally for each node of the network.

- Contention Free Period (CFP): It is composed of Guaranteed Time Slots (GTS) for sensor data, location data and ranging frames transmission, and a GTS request period. Concerning data frames, if source and destination nodes are not physically connected frames are relayed at MAC level using consecutive timeslots. Ranging frames are not relayed and can be sent only between neighbor nodes. Two types of ranging frames are defined: ranging request and ranging response.

- Contention Access Period (CAP): Used for the transmission of command frames through a slotted ALOHA multiple access scheme. Each CAP slot is divided into subslots in order to relay commands.

It should be noted that the relaying procedure is performed at the MAC layer level. When a node has data to transmit, it sends a GTS request on the tree to the coordinator with its address as the source address and the destination address of the transmission. The coordinator, which has the knowledge of the whole network, looks on its routing table if there are relays between the source and the destination. If there are relays, the coordinator determines the route and allocates the GTS for each link.

Functional Architecture and Strategies for Acquisition and Distribution of Location Information

In order to track the position of the target nodes, location information, basically the distances estimated between the target and the anchor nodes, must be acquired and transmitted to a LC that executes the tracking functionality. LCs can be physically located in one or more anchor nodes or in the target nodes. Depending on the location of the LC, several tracking functional architectures (centralized and distributed) can be defined. On the other hand, either the target or the anchor nodes may estimate the distance. This function is referred to as distance acquisition function. The allocation of the distance acquisition function to the target or the anchor nodes is a design alternative that may have an impact on the need of resources.

Tracking Function Distribution in the Network

Depending on the location of the LC function, different tracking functional architectures can be defined. In the tracking functional architecture that we denote as centralized architecture, the tracking functionality is implemented in one or more previously defined anchor nodes that become LCs. Figure 3 shows an example of a centralized architecture with one LC. Using one LC entails a higher need of resources, as multiple hops will be needed to forward the location information to the LC. Defining multiple LCs reduces the need of resources, but increases complexity, as the tracking functionality must be implemented in several nodes and a procedure should be implemented to assign each target to the closest LC.

In the distributed architecture, each target dynamically picks one of its neighbor anchors to execute the tracking functionality. Therefore, there may be as many simultaneous LCs as targets. As the LC is always executed by an anchor neighbor to the target, only one timeslot will be needed to exchange data frames between the target and the LC, and resources will consequently be reduced. As a drawback, the tracking functionality must be implemented in every anchor.

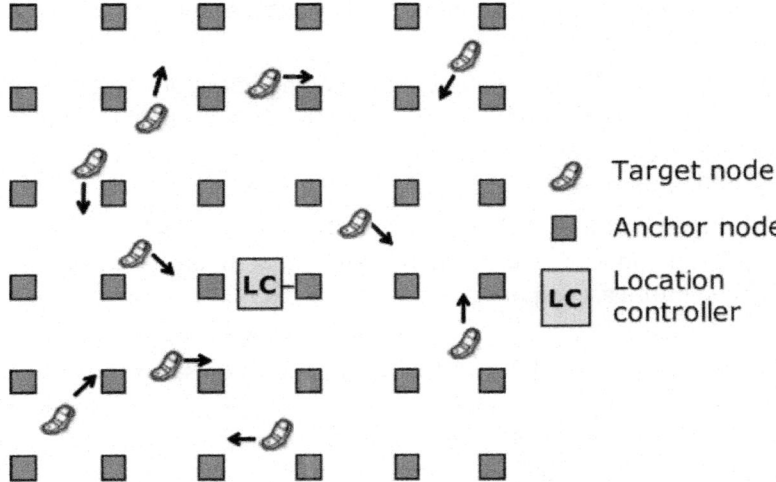

Figure 3: Tracking system. Centralized architecture with 1 LC.

Finally, in the target-centered architecture the LC function is implemented in the target nodes. The target nodes perform ranging with their neighbor anchors and obtain their own position applying the tracking algorithm. Therefore, there is no need of transmitting the estimated distances and the updated position. Nevertheless, the implementation of the tracking functionality requires certain computational capacity on the target nodes, increasing their complexity and cost.

Acquisition and Distribution of Location Information

The acquisition of the location information is done through the ranging procedure. The procedure initiator (target or anchor) transmits a ranging request to another node, which estimates the time of arrival and sends a ranging response after a predefined time. The initiator measures the time of arrival of the response and can estimate the transmission delay and the distance between the nodes (Two Way Ranging). In order to improve the accuracy of distance estimation, two ranging responses can be sent in order to compensate for the clock drift (Three Way Ranging).

Once the distances between the target and the anchor nodes have been estimated, they must be transmitted to the LC in a location data frame. The LC calculates and transmits the updated position to the target node. In order to reduce the amount of resources needed to acquire and distribute the location information, different enhancements can be applied [18].

- Data aggregation: All the distances estimated can be aggregated in a single location data frame and sent to the LC by the ranging initiator (target or anchor).

- Broadcast/multicast request: The ranging initiator (target or anchor) can aggregate multiple ranging requests into a single ranging request sent to all its neighbor nodes (broadcast request) or to a subset of them (multicast request).

- Multicast response: After receiving the ranging requests from different nodes, an anchor or target node can aggregate all the responses into a single multicast response.

If the initiator is the target node, multicast response would require the simultaneous update of the position of all the targets, so the anchors can aggregate the responses to multiple targets into a single multicast response packet. The same applies for broadcast/multicast request and data aggregation when the initiator is the anchor node.

Tracking Function Implementation

With respect to the tracking technique itself, parametric and non-parametric approaches can be distinguished. Parametric approaches compute the location based on the a priori knowledge of a model. On the other hand, non-parametric approaches do not require model knowledge, although in some cases they may use some statistic parameters (mean, variance). Specifically, the following algorithms are considered in this study: Trilateration, Least Square-Multidimensional Scaling (LS-MDS), Least Square-Distance Contraction (LS-DC), Extended Kalman Filter (EKF) and Particle Filter (PF).

Trilateration is a non-parametric algorithm that computes the position based on the distance estimated between the target and three anchor nodes using a geometrical method for determining the intersection of three sphere surfaces [7]. Consequently, regardless of the number of anchors selected, only the three anchors with smallest estimated distance to the target are used for position computation.

The algorithm LS-MDS is a completely non-parametric approach combining Multidimensional Scaling (MDS) with Least Squares (LS) minimization [19]. MDS is a multivariate data analysis technique used to map "proximities" into a space. These "proximities" can be either dissimilarities (distance-like quantities) or similarities (inversely related to distances). Given n points and corresponding dissimilarity, MDS finds a set of points in a space such that a one-to-one mapping between the original configuration and the reconstructed one exists. Then it is possible to map back the solution to the absolute reference system by Procrustes transformation. MDS is used to obtain a previous estimation of the solution. Then, the localization problem is posed into a non linear least squares optimization problem. The goal is to obtain the matrix of computed positions X that minimizes the stress function σ (X) defined as follows:

$$\sigma(X) = \sum_{i=1}^{n} \sum_{j=1}^{n} (\partial_{ij} - d_{ij}(X))^2 \tag{1}$$

where ∂_{ij} is the estimated distance between nodes i and j and $d_{ij}(X)$ is the distance between nodes i and j associated with the computed node locations X. In order to solve this problem, a low-complexity algorithm based on majorization technique is applied. Specifically, the algorithm is known as SMACOF and it consists of an iterative procedure that attempts to find the minimum of a non-convex function by tracking the global minima of the so-called majored convex function successively constructed from the original objective and basis on the previous solution.

LS-DC combines the Distance Contraction (DC) algorithm with Least Square minimization [20]. Firstly, the distances between the target and n anchor nodes are estimated. Each estimated distance defines a line of position for the target's location as a circle around the corresponding anchor node. Then, the feasibility region is defined as the area of intersection between the n circles. If the feasibility region does not exist, distance contraction cannot be applied and the LS-MDS approach is used instead. If the feasibility region exists, an initial point is computed inside the feasibility region and the contracted distances are computed as the shortest distance from each anchor to the aforementioned feasibility region. Finally, a minimization algorithm is applied using the contracted distances instead of the estimated ones. Since the function becomes convex, any minimization algorithm (*i.e.*, global distance continuation, steepest descent) can be used thus reducing complexity, although here SMACOF has been used in order to be comparable with LS-MDS.

The Extended Kalman Filter is a Bayesian technique known for its low complexity, performance and stability as a tracking algorithm [21]. EKF addresses the problem of trying to estimate the state x of a discrete-time controlled process that is governed by a non-linear stochastic difference equation:

$$x_t = f\left(x_{t-1}, u_{t-1}, w_{t-1}\right)$$

(2)

with a measurement z that is:

$$z_t = h\left(x_t, e_t\right)$$

(3)

The non-linear function f in Equation (2) relates the state at the previous time step $t-1$ to the state at the current time step t and includes as parameters any driving function u_t and the process noise w_t. The non-linear function h in Equation (3) relates the state x_t to the measurement z_t and includes as parameter the measurement noise e_t. Process and measurement noise are assumed to be independent, white, and with normal probability distributions $p(w) \in N(0, Q)$ and $p(e) \in N(0, R)$. The process and measurement noise covariance matrixes Q and R are defined by variances σ_w^2 and σ_e^2.

The Kalman-based tracking algorithm has two major stages, namely, the update and the correction stages, which are iterated k times for every observation occurring at a given time. The time update equations project the state and covariance estimates from the previous time step $t-1$ to the current time step t. The measurement update equations correct the state and covariance estimates with the measurement z_t. As f and h cannot be applied to the covariance directly, matrixes of partial derivatives (Jacobian) are computed.

Focusing on the implementation of the EKF for the tracking application, the state vector x contains target's position p_t and speed v_t as process variables. The measure vector z contains the process observations, namely the estimated distances between the target and the anchors. The functions that describe the evolution of the state vector through time and the relation between the state vector and the measure vector are:

$$\begin{pmatrix} p_{t+1} \\ v_{t+1} \end{pmatrix} = \begin{pmatrix} I & T_s \cdot I \\ 0 & I \end{pmatrix} \begin{pmatrix} p_t \\ v_t \end{pmatrix} + \begin{pmatrix} T_s^2 / 2 \cdot I \\ T_s \cdot I \end{pmatrix} w_t$$

(4)

$$\tilde{z}_t(i) = \mid p_i - \tilde{p}_t \mid + e_t$$

(5)

where T_s is the time between two consecutive updates and p_i is the position of anchor i.

Finally, Particle Filters are recursive implementations of Monte Carlo based statistical signal processing. The use of particle filters for positioning in wireless networks was proposed in [22]. The particle filter is based on a high number of samples of the state vector or particles, which are weighted according to their importance (likelihood) in order to provide an estimation of the state vector. The advantage of particle filters over other parametric solutions is that non-linear models and non-Gaussian noise can be defined. As a drawback, their computational complexity is higher, so they are suitable in applications where computational power is rather cheap and the sampling rate slow. As for EKF, a state vector x, a measure vector z and functions f and h are defined. On each step, the particles are moved according to Equation (4) and the weights are updated according to the likelihood of the observations:

$$w_t^i = w_{t-1}^i p(z_t \mid x_t^i), \ i = 1, ..., N$$

(6)

where i is the particle index, N is the number of particles and the probability p(zt|xit) is equivalent to the probability pe(zt−h(xit)) according to the distribution of the measurement error e. But here the measurement noise e is not necessarily considered Gaussian. Specifically, we have defined the measurement error model as a weighted sum of three Gaussian components for the different channel configurations (LOS/NLOS/NLOS2). The weight of each component is also a Gaussian-like function as will be later defined in Equation (9). Consequently, the filter is defined by the variance of process noise $_w^2$ and the parameters of the measurement error model (mean and variance of each component and mean and variance of each weight).

PERFORMANCE EVALUATION

System Model and Simulator Description

In order to evaluate the impact of the different system design alternatives and parameters, we have developed a specific simulation application using C++. The simulation scenario represents a relatively wide indoors area, such as a warehouse, where people and goods moving at pedestrian speeds will be tracked with accuracy below 1 m. On this scenario a UWB-based wireless sensor network composed of N_a anchor nodes and N_m target nodes is deployed. The existence of walls and obstacles is considered through the use of an indoor ranging model which accounts for the probability of non-line-of-sight between targets and anchors. The dynamics of the targets are modeled by a Random Walk Model, with random directions and speeds that are constant during a certain period of time, after which new random directions and speeds are set [23].

A common set of parameters has been defined. Area size has been set to 50 m × 50 m in order to represent a relatively wide indoors area. UWB nodes range has been set to 15 m according to the specifications of the IR-UWB platform presented in Section 3.1 [14]. In order to guarantee connectivity between adjacent anchors, the distance between adjacent anchors has been set to 10 m, which results into 36 anchors. Concerning the dynamics of the target nodes, the Random Walk Model is defined by the minimum speed, the maximum speed and the change interval, that have been set to 0.1 m/s, 3 m/s and 20 s respectively in order to model pedestrian motion. Finally, as a result of prior simulations, the nominal position update interval has been set to 976 ms (5 superframes), as it provides accurate tracking of the moving targets with a reasonable use of resources (timeslots).

A ranging model is used to characterize the ranging error distribution and to generate the distance estimation samples. Range measurements based on round-trip TOA estimation through n-Way Ranging transactions can be modeled as:

$$\tilde{d}_{ij} = d_{ij} + \varepsilon_{ij} + n_{ij} = d'_{ij} + n_{ij} \tag{7}$$

where d_{ij} is the actual distance between nodes i and j, d_{ij}' is the biased distance (with bias ε_{ij}) and n_{ij} is a residual error term.

The biased distance is modeled as a weighted sum of Gaussian and Exponential components conditioned upon the actual distance and channel configuration. The pdf of d', conditioned upon d and a particular channel configuration C, is described as follows:

$$p_C[d'/(d,C)] = G_C \frac{1}{d} \frac{1}{\sqrt{2\pi}\sigma_C} e^{-\frac{\left(\frac{d'}{d}-1\right)^2}{2\sigma_C^2}} + E_C \frac{1_{\{d'>d\}}}{d} \lambda_C e^{-\lambda_C\left(\frac{d'}{d}-1\right)} \tag{8}$$

where $d \neq 0$, $\{G_C, \sigma_C\}$ and $\{E_C, \lambda_C\}$ are the weights and parameters of Gaussian and Exponential mixture components, $1_{\{x>y\}} = 1$ whenever $x > y$ and 0 otherwise, and C takes its value among {LOS, NLOS, NLOS2}. The model is enhanced by taking into account the probability $W_C(d)$ to have a particular channel configuration at a distance d. These weights are described as Gaussian-like functions:

$$W_C(d) = \frac{\xi}{\sqrt{2\pi}\varsigma_C} e^{-\frac{(d-d_C)^2}{2\varsigma_C^2}} \tag{9}$$

This model for the ranging bias was proposed and validated through a measurement campaign with real UWB equipment in an office environment in [24], where the values for the different parameters of the model were also identified.

The residual error is modeled as additive and centered, with a variance σ_n^2 that depends on detection error terms affecting unitary TOA estimates, *i.e.*, receiver sampling rate, and involved protocol durations, and is independent of the distance between the nodes.

In order to set a realistic value for ranging residual error σ_n, a measurement campaign was carried out using the open IR-UWB platforms mentioned in Section 3.1. Line-of-sight and distances up to 5 m in steps of 0.25 m were considered. Under these conditions, the error is mostly due to residual error. For each distance, 150 ranging samples were obtained and the mean and standard deviation of the distance estimation error was computed. As it can be observed in Figure 4, the mean is always close to zero and no dependency on distance has been detected, as it was expected for ranging residual error. On the other hand, standard deviation varied from 10 cm to 40 cm, with an average value of 25 cm. As a result, we will consider a value of $\sigma_n = 0.3$ m for the subsequent simulations.

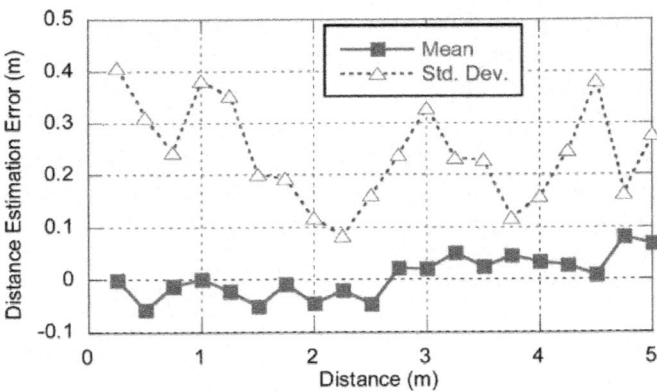

Figure 4: Distance estimation error.

Simulation Results

In this section, the system performance is evaluated and the different design alternatives are assessed. In Subsection 4.2.1 the performance of the different LT algorithms considered is evaluated using an ideal no-delay approach. The effect of delay on LT algorithms is analyzed in Subsection 4.2.2 considering generic position update latency. In Subsection 4.2.3 the specific MAC structure is taken into account, and the effect of target mobility is discussed. In Subsection 4.2.4 the different proposed strategies for acquisition and distribution of location information are assessed with the common centralized architecture with 1 LC, followed by the assessment of more advanced tracking functional architectures in subsection 4.2.5. The complete system evaluation considering all the different aspects is provided in subsection 4.2.6. Finally, experimental results with UWB prototypes are also provided.

Tracking Algorithms

The performance of each algorithm has been evaluated depending on the number of anchors used for positioning. In a first step, distance estimation and position calculation are ideally considered instantaneous, so the potential error associated with delay is not present. Concerning EKF and PF, prior simulations have been carried out in order to find the optimum values of the process and measurement noise parameters that minimized the error.

Figure 5 shows the average positioning error for a ranging residual error $_n$ = 0.3 m. As it can be observed, the best performance is achieved with the particle filter. Nevertheless, the performance of the particle filter is far better than it can be expected in a real situation. The reason is that, after being optimized through simulations, the measurement error model of PF behaves almost exactly as the ranging model used to generate the distance estimation samples. Consequently, PF can deal even with highly biased measurements and the error of the particle filter decreases as the number of anchors used for location increases. In a real system, the precise characterization of the specific ranging model of the scenario would require costly measurement and calibration phases, and the use of a generic model would not provide so good results.

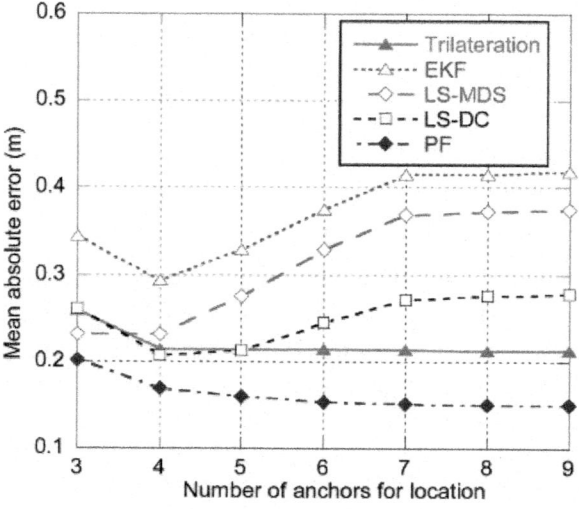

Figure 5: Positioning error. Distance between anchors = 10 m, σ_n = 0.3 m.

Trilateration, LS-MDS and LS-DC show a similar value for minimum error slightly over 20 cm, compared to 30 cm for EKF. The optimum number of anchors is four for LS-DC and EKF and three for LS-MDS. If more anchors are used, the added anchors will be more distant and will have higher ranging bias, thus increasing the positioning error. On the other hand, trilateration is almost independent on the number of anchors used to compute the position, as only the three closest anchors will be used. For every algorithm, there is an increase of the error when only three anchors are used, and the error remains constant for more than seven anchors, as the target is not likely to be in coverage of more than seven anchors.

Figure 6 shows the average positioning error for a more pessimistic value of ranging residual error ($\sigma_n = 0.6$ m). As expected, the average error is increased compared to the same configuration with ranging residual error $\sigma_n = 0.3$ m. The error increase for trilateration is especially remarkable, with results comparable to EKF in terms of minimum error. This means that trilateration requires accurate TOA estimation in order to provide good results, as it always uses three measurements for position computation and cannot take advance of diversity of measurements. Consequently, LS-MDS and LS-DC are preferred over trilateration.

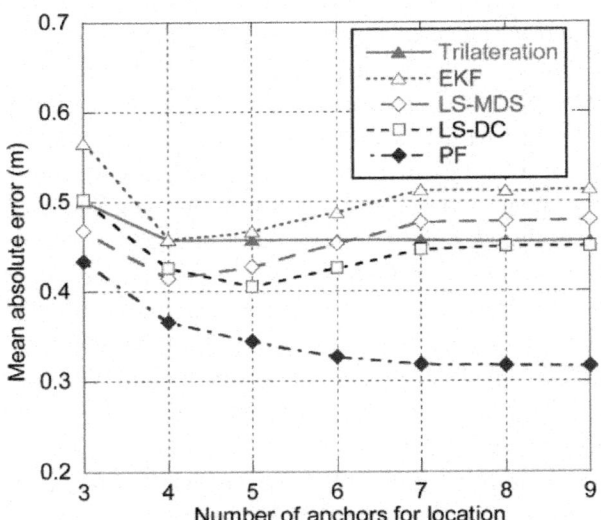

Figure 6: Positioning error. Distance between anchors = 10 m, σ_n =0.6 m.

As it was previously mentioned, parametric approaches such as EKF and PF require the accurate characterization of the target's motion model and the measurement model. With this purpose, the optimum values of the different parameters of EKF and PF that minimized the error for certain conditions (scenario layout and ranging error) were obtained through prior simulations. In order to assess the impact of the accuracy of model characterization on the positioning error, the performance of EKF and PF using parameter values that are not optimized for the current conditions is compared to the performance of the optimized filters. Figure 7 shows the average positioning error for both optimized and non-optimized EKF and PF for a ranging residual error σ_n = 0.3 m. Optimized filters use the parameter values resulting of the optimization for σ_n = 0.3 m as in Figure 5. On the other hand, in order to assess the impact of an incorrect calibration, non-optimized filters use parameter values different from the optimum. In case of the non-optimized EKF, instead of using the optimum values (resulting of the optimization for σ_n = 0.3 m), the measurement noise variance σ_e^2 is set to the value resulting of the optimization for σ_n = 0.6 m. That is to say, we use a calibration obtained for different conditions (σ_n = 0.6 m) instead of the calibration obtained for the current conditions (σ_n = 0.3 m). As it can be observed, there is little degradation in the EKF performance, but it should be noted that the optimized EKF showed the worst performance among the different algorithms, as the Gaussian measurement noise assumption is not appropriate to model the ranging error, which is highly biased. Concerning the non-optimized PF, the value of the variance of the LOS component optimized for $_n$ = 0.6 m has been used instead of the value optimized for $_n$ = 0.3 m, and the mean and variance of the NLOS and NLOS2 components have been multiplied by factors 2 and 4 respectively. As expected, the non-optimized PF shows worse performance than the optimized PF, with a minimum error slightly over 20 cm, which is comparable to the performance achieved with non-parametric approaches such as LS-DC (see Figure 5). It must be noted that the parameters (mean and variance) defining the weights of each component have not been modified, although these values depend on the specific scenario and would introduce additional degradation. Finally, it should be also remarked that the difference between EKF and PF performance is mostly due to the fact that EKF uses a simple Gaussian measurement model whereas in PF we have implemented a 3-component model that is able to deal with the biased ranging error, rather than by the filters themselves. Although more complex models can be implemented by combining multiple EKF filters through multihypothesis tracking and Interacting Multiple Model (IMM) methods, the PF is better suited for the implementation of complex measurement models.

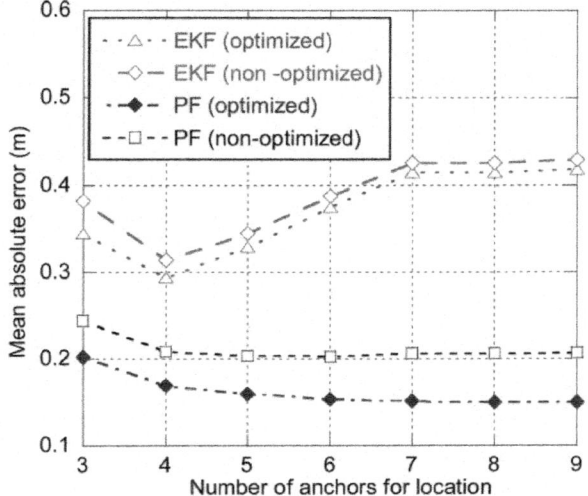

Figure 7: Impact of model optimization in EKF and PF. Distance between anchors = 10 m, σ_n = 0.3 m.

Effect of Position Update Latency

According to the position update process previously described and to the MAC superframe structure shown in Section 3.1, there are many sources of delay in the position update process: delay associated to the request and allocation of free ranging slots, duration of the ranging exchanges, transmission of the estimated distances to the location controller, latency of position computation, transmission of the updated position to the target, *etc.*

Delay prior to the start of the ranging exchanges has no appreciable effect on positioning accuracy. Nevertheless, delay between the ranging exchanges and the final position availability has a negative effect on positioning accuracy due to the movement of the target. We define position update latency as the time between the start of the ranging exchanges and the availability of the position at the target. Figure 8 shows the effect of position update latency on the average positioning error for the different tracking algorithms. As it can be observed, all the algorithms show a similar evolution and the error grows as position update latency increases. For 200 ms that is approximately the duration of a MAC superframe, the average error increases around 18 cm. For 400 ms (two superframes) the error increase can be as high as 45 cm. Therefore, position updates should be carried out preferably within a single superframe.

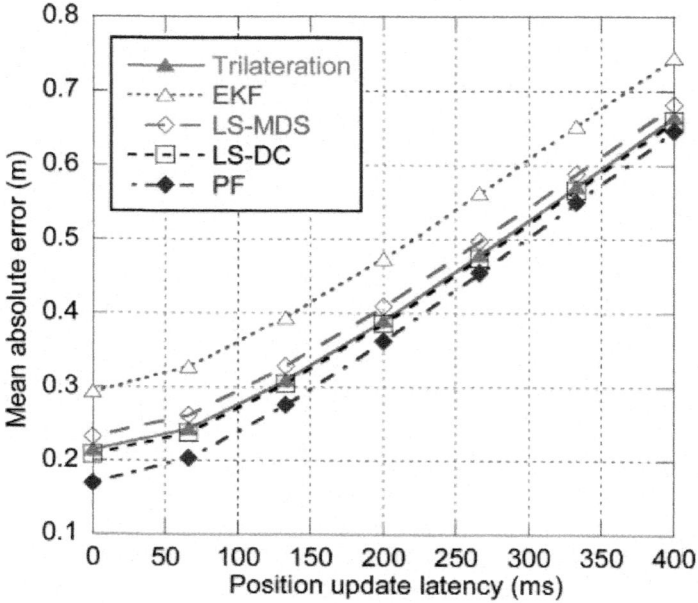

Figure 8: Positioning error depending on position update latency.

Effect of Target Mobility

Next, the effect of target mobility on the positioning error is analyzed, taking into account the timing associated to the MAC superframe structure presented in Section 3.1. Figure 9 shows the positioning error for the centralized architecture with a single LC and four anchors used for location. As it can be observed, when delays are taken into account, all the algorithms are degraded due to the movement of the target between distance estimation and position computation. Non-parametric algorithms, namely trilateration, LS-MDS and LS-DC, which are ideally independent on target speed, have a similar evolution. EKF severely degrades for speeds greater than 1.5 m/s as the estimation is based on target's previous position. Finally, although PF uses the target's dynamic model to move the particles, particles are weighted on each step according to their likelihood, so the new position is almost independent of the previous one and degradation is slightly higher than for non-parametric methods.

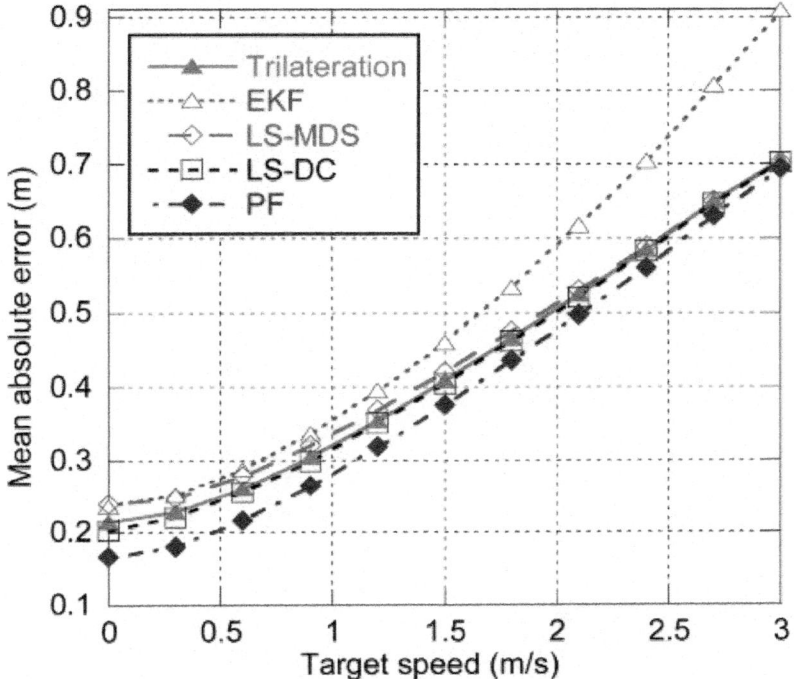

Figure 9: Positioning error depending on target speed.

Impact of Acquisition and Distribution Strategies

In order to reduce the amount of timeslots needed and consequently the latency, data acquisition and distribution enhancements presented in Section 3.2.2 can be applied. The following notation is used: SRq (Single Request), MRq (Multicast Request), SRp (Single Response), MRp (Multicast Response), NDA (No Data Aggregation), DA (Data Aggregation). The different enhancements have been simulated based on a centralized architecture with one location controller (denoted as 1LC in Figure 10) and one target. Trilateration has been used as location and tracking algorithm as its performance, without considering latency, is almost independent of the number of anchors used for location provided that more than three anchors are used (see Figure 5). This way, when the amount of timeslots needed and consequently the latency linked to data acquisition and distribution is explicitly considered, the impact of each strategy depending on the number of anchors can be appreciated better. As Three-Way ranging is used, two ranging responses are generated for each request, so three slots are needed for each ranging exchange. Concerning estimated distances transmission, in general multiple hops will be needed to relay the data frames from the target to the LC and *vice versa*. As it was specified in Section 3.1, there are 12 ranging slots and eight data slots per superframe.

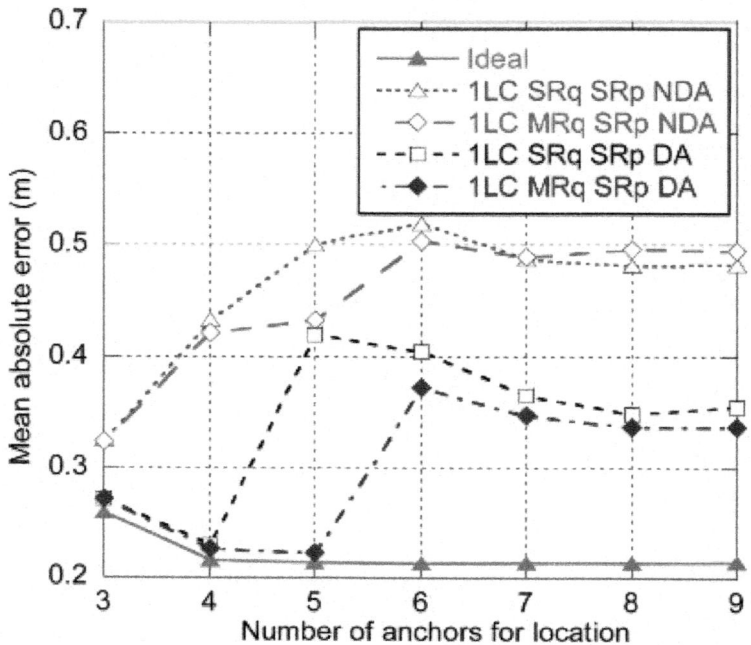

Figure 10: Positioning error for the enhanced modes.

Figure 10 shows the average positioning error for each one of the acquisition & distribution schemes depending on the number of anchors. The error for an ideal case with no delay is included as a reference. Position update latency and therefore positioning error increase are mainly determined by the number of superframes needed for ranging exchanges and estimated distances transmission. Note that results are constant for seven or more anchors as with this configuration (10 m between anchors) the target is not likely to be in coverage of more than seven anchors. The modes without data aggregation show a similar evolution as they are mainly determined by the number of superframes needed for estimated distances transmission, which depends on the number of anchors used and the number of hops from the target to the location controller. Consequently the error increases as the number of anchors increase. The only difference is for five anchors, as with SRq at least two superframes are needed for ranging exchanges, whereas with MRq ranging exchanges can be completed within one superframe and a second superframe may not be needed if the target is just one hop away from the LC.

When data aggregation is used, position update latency is mainly determined by the ranging exchanges, as location data transmission will be carried out always in a single superframe. With single request (SRq SRp DA), a second superframe will be needed for ranging when five anchors are used, so there is an important increase of the error. Then the error slightly decreases, which is related to the ratio of distances recently estimated (in the second superframe) and delayed (in the first superframe). With multicast request (MRq SRp DA) the error increase occurs when six anchors are used.

Next the effect of tracking multiple targets simultaneously on system capacity is analyzed. With that purpose, the number of anchors used for location is fixed to four and the number of targets is variable. Figure 11 shows the % of GTS slots used for location data transmission. When no enhancements are applied, the amount of slots used increases quickly and the system is eventually saturated for more than four targets, with a residual 20% left for sensor data communication. The amount of GTS slots used is reduced for the enhanced modes. As with DA a single measurement report packet is sent, capacity of these modes is limited by the availability of ranging slots to five targets (SRq SRp DA), six targets (MRq SRp DA) and eight targets (MRq MRp DA) per picocell.

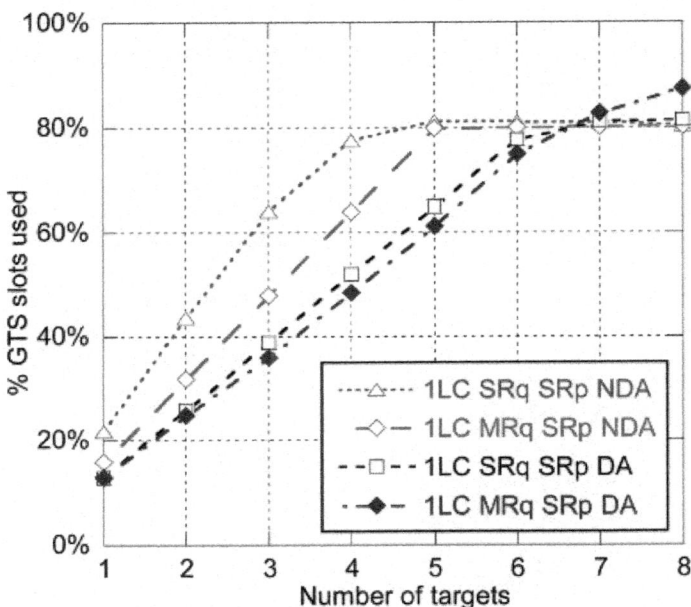

Figure 11: % of GTS slots used for location data transmission for the enhanced modes.

Impact of the Tracking Functional Architecture

In this section, the different functional architectures of the tracking system that were proposed in Section 3.2.1 are evaluated. Figure 12 shows the average error obtained for the different functional architectures depending on the number of anchors used for location. SRq SRp NDA mode has been considered. Results for the centralized architecture with one LC were already discussed in the previous section. For the distributed architecture, as the LC is implemented in an anchor neighbor to the target, location data transmission can always be done within a superframe, and latency is not determined by location data transmission, but by ranging exchanges. Specifically, as 12 ranging slots are available on each superframe and three slots are required for each ranging exchange in the SRq SRp NDA mode, up to four ranging exchanges can be done in a single superframe, and a second superframe will be needed if more than four anchors are used. Consequently there is an important increase of the error when five anchors are used, and a slight decrease for more anchors, which is again related to the ratio of distances recently estimated (in the second superframe) and delayed (in the first superframe). A similar evolution is shown for a centralized architecture with four LC as most of the time the target will be neighbor to one of the LCs. Finally, the target-centered architecture shows a similar evolution but with a slightly lower error, as there is no need of transmitting the estimated distances and the computed position, so the position update latency will always be a little bit lower.

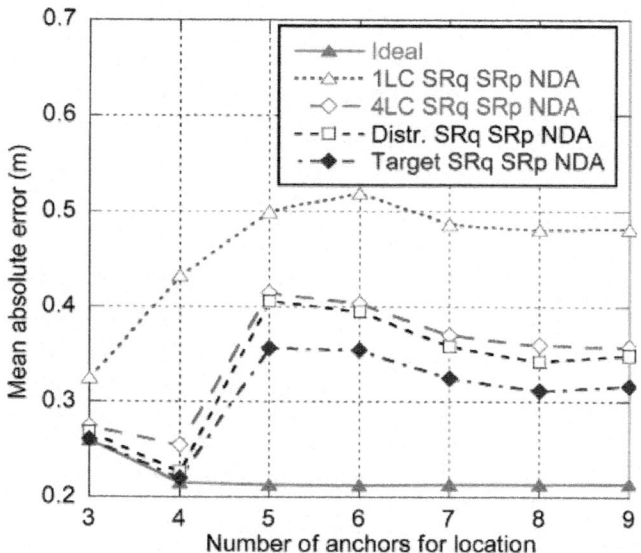

Figure 12: Positioning error for the tracking functional architectures (SRq SRp NDA).

Figure 13 shows the % of GTS slots used for location data transmission depending on the number of targets with 4 anchors used for location. As it was previously mentioned, the centralized architecture with one LC can track up to four targets until slots allocated to location data transmission are saturated. The distributed architecture and the centralized architecture with four LCs can track up to five targets until ranging slots are saturated, with a residual 10–15% of slots available for sensor data transmission. Finally the target-centered architecture can also track up to five targets but, as location data transmission is not needed, has a lower use of resources, and consequently higher capacity available for sensor data transmission. Therefore, the target-centered architecture is the optimal in terms of use of resources.

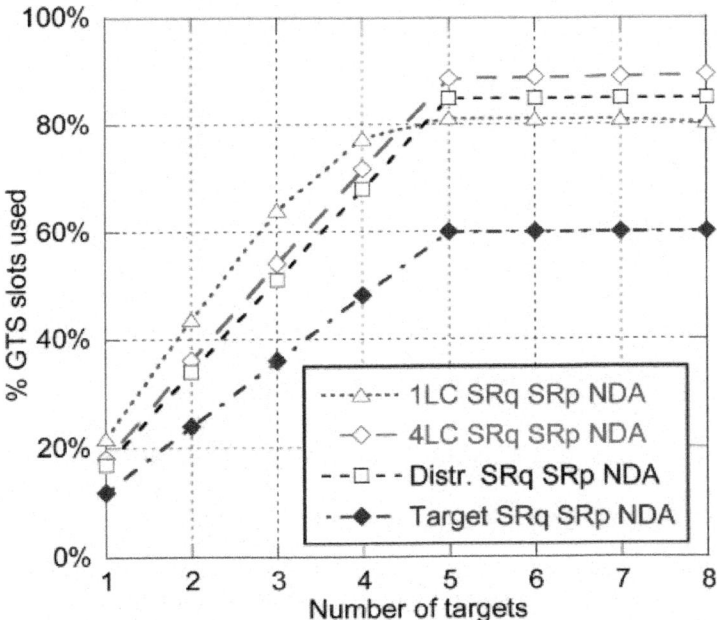

Figure 13: % of GTS slots used for location data transmission for the different architectures (SRq SRp NDA).

Complete System Evaluation

Finally, accuracy and capacity of the system are evaluated for the different algorithms considering the real MAC implementation. In order to minimize latency, the target-centered architecture centralized has been considered, together with MRq, SRp and DA enhancements. MRq has not been considered as it requires the coordination of the updates, which is complex for the target-centered architecture. Figure 14 shows the average positioning error for the different algorithms. Results for 5 or less anchors are similar than those of Figure 5 as latency is relatively small. When 6 anchors are used, there is an important increase of the error for all the algorithms, as each update will require two superframes thus increasing latency.

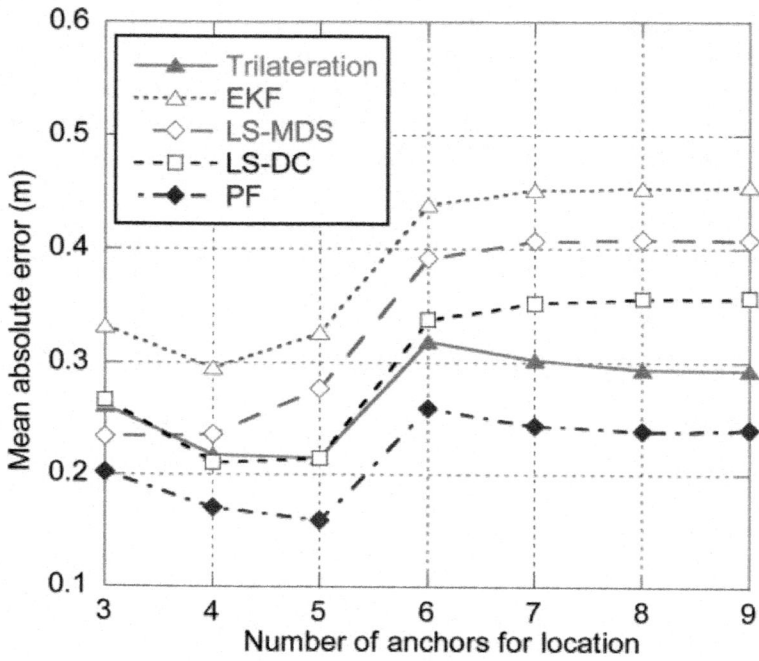

Figure 14: Positioning error for the target-centered architecture (MRq SRp DA).

In order to evaluate capacity, we have selected two options, LS-MDS with three anchors and LS-DC with four anchors, which provide an average positioning error of 23 and 21 cm respectively. Although PF provides better performance, non-parametric approaches such as LS-DC and LS-MDS are preferred over PF as they are completely independent of the scenario and do not require any prior model characterization. Again, the target-centered architecture and MRq, SRp and DA enhancements have been considered. As it can be observed in Figure 15, the system can track up to six anchors in case of LS-DC with four anchors, and up to eight anchors in case of LS-MDS with three anchors, leaving a residual 40% of GTS slots available for sensor data transmission.

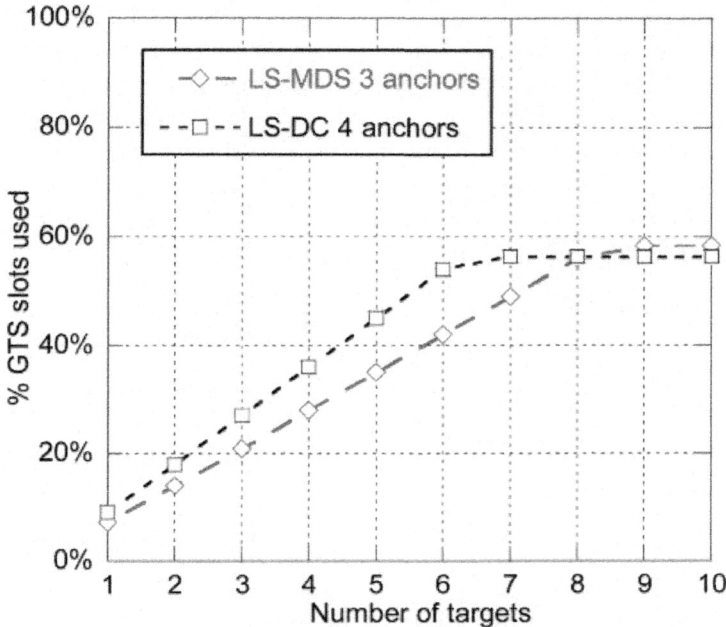

Figure 15: % of GTS slots used for location data transmission for LS-MDS (three anchors) and LS-DC (four anchors).

Experimental Results

In order to assess the accuracy level that can be reached with real UWB equipment, an experiment was carried out in a room with approximate dimensions of 5 × 3 meters using the open IR-UWB platforms already introduced in Section 3.1. These platforms are prototypes including radio-frequency, baseband and MAC hardware boards and a software MAC running on an FPGA [14]. The modulation scheme is based on Differential Binary Phase Shift Keying (DBPSK), while demodulation is performed using differential correlation between the incoming signal corresponding to the current data symbol and the previous one. The MAC layer is based on IEEE 802.15.4 and has been already explained on Section 3.1, while the main PHY/ MAC parameters are shown in Table 1.

Five prototype devices have been used, one configured as picocell coordinator and anchor node, another one as target node, and the other three as anchor nodes. As the devices are started, the target node estimates the distances to each anchor node and sends them to the picocell coordinator. The picocell coordinator transmits the estimated distances through a serial RS-232 interface. Finally, a computer connected to the picocell coordinator retrieves the distances and computes the position. Consequently, the tracking functional architecture is centralized with one LC, and four anchors are used to locate a single target. LS-DC was implemented as location algorithm as it provides good performance on the different configurations simulated, is independent on target speed and is completely non-parametric, so no prior model characterization is required.Figure 16 shows a plan of the measurement scenario. The anchor nodes (green dots) were placed near the corners of the room and the estimated position was measured in 13 different locations (blue dots).

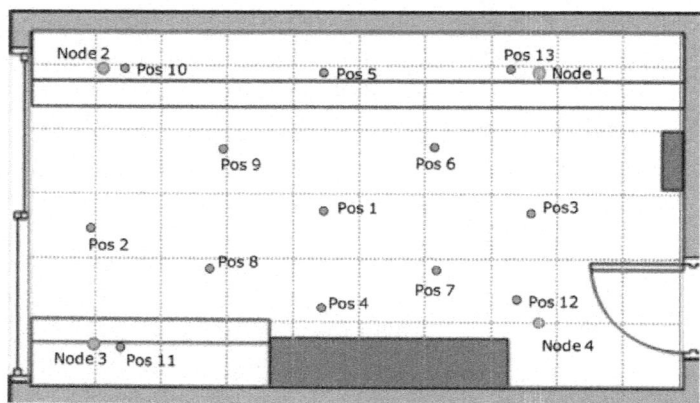

Figure 16: Measurement scenario.

Table 2 shows the mean and standard deviation of the positioning error on each one of the positions surveyed. The mean absolute error (MAE) varies from 7.8 cm when the target is in the middle of the area (position 1) to 40–45 cm when the target is in a corner of the area (positions 10, 11, 12 and 13). Standard deviation also increases for the corner placements. Average MAE is 26.6 cm, which is slightly higher than the 21 cm obtained for LS-DC in the simulations.

Table 2: Positioning error for the different positions surveyed

Positioning error	Pos1	Pos2	Pos3	Pos4	Pos5	Pos6	Pos7	Pos8	Pos9	Pos10	Pos11	Pos12	Pos13
Mean absolute error (cm)	7.8	41.6	29.4	13.6	19.4	14	15.8	17.9	20.5	44.3	44.9	37	40.2
Std deviation (cm)	2.6	4.1	2.8	6.1	2.3	5.3	7.3	3.2	10.9	11	11.8	12.5	5.1

CONCLUSIONS

In this paper, a UWB-based communication and location tracking system for Wireless Sensor Networks has been analyzed. Besides distance estimation and position calculation algorithms, the study covers some other aspects not usually considered on existing studies, such as position update latency, mobility, the functional architecture, the transmission of the associated information, and the need of tracking multiple terminals.

Concerning location and tracking algorithms, the particle filter provides the best results, although measurement model characterization would entail a costly calibration phase, and the use of a generic model would not provide so good results. LS-MDS and LS-DC provide good results and do not require model characterization. Finally, trilateration shows good results for accurate ranging measurements, but degrades as ranging error increases, and the Extended Kalman filter only works well for slow moving targets but severely degrades as target speed increases.

But, as it has been shown, positioning error is highly sensitive to position update latency. Furthermore, when combined data communication and location systems are considered, the use of temporal resources is critical, as it has an impact both on the capacity of the tracking system and the sensor data transmission capacity. In order to deal with latency and resource limitation, different solutions are proposed. The target-centered architecture is optimal in terms of latency and resources needed, as there is no need of transmitting the estimated distances to the network. For any other tracking functional architecture, the use of data aggregation is essential in order to minimize the amount of slots used for location data transmission. Finally, the number of anchors used to locate a certain target should be limited according to the number of slots allocated to ranging, so the position update can be completed within a single superframe.

ACKNOWLEDGMENTS

This work was supported by the European research project EUWB which is partly funded by the Commission of the European Union under the 7th European Framework Programme for Research and Technological Development (FP7) and here under the Information and Communication Technologies (ICT) research programme and by Gobierno de Aragón for WALQA Technology Park.

REFERENCES

1. Liu, H; Darabi, H; Banerjee, P; Liu, J. Survey of wireless indoor positioning techniques and systems. *IEEE Trans. Syst. Man Cybern. Part C* 2007, *37*, 1067–1080.

2. Gu, Y; Lo, A; Niemegeers, I. A survey of indoor positioning systems for wireless personal networks. *IEEE Commun. Surv. Tutor* 2009, *11*, 13–32.

3. Yang, L; Giannakis, GB. Ultra-Wideband communications, an idea whose time has come. *IEEE Signal Process. Mag* 2004, *21*, 26–54.

4. Gezici, S; Tian, Z; Giannakis, GB; Kobayashi, H; Molisch, AF; Poor, HV; Sahinoğlu, Z. Localization via ultra-wideband radios. A look at positioning aspects of future sensor networks. *IEEE Signal Process. Mag* 2005, *22*, 70–84.

5. Dardari, D; Conti, A; Ferner, UJ; Giorgetti, A; Win, MZ. *Ranging with Ultrawide Bandwidth Signals in Multipath Environments*; Institute of Electrical and Electronics Engineers: Washington, DC, USA, 2009; Volume 97, pp. 404–426.

6. Güvenç, I; Sahinoğlu, Z; Orlik, P. TOA estimation for IR-UWB systems with different transceiver types. *IEEE Trans. Microw. Theory Tech* 2006, *54*, 1876–1886.

7. Sahinoglu, Z; Gezici, S; Güvenc, I. *Ultra-Wideband Positioning Systems*; Cambridge University Press: Cambridge, UK, 2008; pp. 63–100.

8. Yu, K; Montillet, JP; Rabbachin, A; Cheong, P; Oppermann, I. UWB location and tracking for wireless embedded networks. *Signal Process. Spec. Section Signal Process. UWB Commun* 2006, *86*, 2153–2171.

9. IEEE 802.15 Working Group. *Part 15.4: Wireless Medium Access Control (MAC) and Physical Layer (PHY) Specifications for Low-Rate Wireless Personal Area Networks (LR-WPANs)*, Available online: http://profsite.um.ac.ir/~hyaghmae/ACN/WSNMAC1.pdf (accessed on 10 August 2011).

10. IEEE 802.15 Working Group. *Wireless Medium Access Control (MAC) and Physical Layer (PHY) Specifications for Low-Rate Wireless Personal Area Networks (WPANs): Alternate PHYs*, Available online: http://www.it-expo.org/docs/9_2_en.pdf(accessed on 10 August 2011).

11. Chu, Y; Ganz, A. A UWB-based 3D location system for indoor environments. Proceedings of the 2nd International Conference on Broadband Networks (BroadNets 2005), Boston, MA, USA, 3–7 October 2005; pp. 1147–1155.

12. Lo, A; Xia, L; Niemegeers, I; Bauge, T; Russell, M; Harmer, D. Europcom—An ultra-wideband (UWB)-based *ad hoc* network for emergency applications. Proceedings of the 67th IEEE Vehicular Technology Conference (VTC2008-Spring), Marina Bay, Singapore, 11–14 May 2008; pp. 6–10.

13. Wu, D; Bao, L; Li, R. UWB-based localization in wireless sensor networks. *Int. J. Commun. Network Syst. Sci* 2009, *2*, 407–421.

14. Pezzin, M; Bucaille, I; Schulze, T; Pato, AV; de Celis, L. An open IR-UWB platform for LDR-LT applications prototyping. Proceedings of the 6th IEEE Workshop on Positioning, Navigation and Communication (WPNC'09), Hannover, Germany, March 2009; pp. 285–293.

15. Giancola, G; Blazevic, L; Bucaille, I; de Nardis, L; di Benedetto, MG; Durand, Y; Froc, G; Cuezva, BM; Pierrot, J; Pirinen, P; *et al.* UWB MAC and network solutions for low data rate with location and tracking applications. Proceedings of the 2005 IEEE International Conference on UWB (ICUWB 2005), Zurich, Switzerland, 5–8 September 2005; pp. 758–763.

16. Venkatesh, S; Buehrer, RM. Multiple-access design for *ad hoc* UWB position-location networks. Proceedings of IEEE Wireless Communications and Networking Conference (WCNC 2006), Las Vegas, NV, USA, 3–6 April 2006; pp. 1866–1873.

17. Bucaille, I; Tonnerre, A; Ouvry, L; Denis, B. MAC layer design for UWB low data rate systems: PULSERS proposal. Proceedings of the 4th IEEE Workshop on Positioning, Navigation and Communication (WPNC'07), Hannover, Germany, March 2007; pp. 277–283.

18. Macagnano, D; Destino, G; Esposito, F; Abreu, GTF. MAC performances for localization and tracking in wireless sensor networks. Proceedings of the 4th IEEE Workshop on Positioning, Navigation and Communication (WPNC'07), Hannover, Germany, March 2007; pp. 297–302.

19. Macagnano, D; Abreu, GTF. Tracking multiple targets with multidimensional scaling. Proceedings of the 9th International Symposium on Wireless Personal Multimedia Communications (WPMC 2006), San Diego, CA, USA, 17–20 September 2006; pp. 1118–1123.

20. Destino, G; Abreu, G. Improving source localization in NLOS conditions via ranging contraction. Proceedings of the 7th Workshop on Positioning, Navigation and Communication (WPNC'10), Dresden, Germany, 7–8 April 2010; pp. 56–61.

21. Daum, F. Nonlinear filters: Beyond the Kalman filter. *IEEE Aerosp. Electron. Syst. Mag* 2005, *20*, 57–69.

22. Gustafsson, F; Gunnarsson, F; Bergman, N; Forssell, U; Jansson, J; Karlsson, R; Nordlund, PJ. Particle filters for positioning, navigation and tracking. *IEEE Trans. Signal Process* 2007, *50*, 425–437.

23. Bai, F; Helmy, A. A survey of mobility models in wireless *ad hoc* networks. In *Wireless* Ad Hoc *and Sensor Networks*, 1st ed; Safwat, A, Ed.; Springer Verlag: Berlin, Germany, 2007.

24.   Denis, B; Pierrot, JB; Abou-Rjeily, C. Joint distributed time synchronization and positioning in UWB *ad hoc* networks using TOA. *IEEE Trans. Microw. Theory Tech. Spec. Issue Ultra Wideband* 2006, *54*, 1896–1911.

Chapter 7

DESIGN OF CMOS INTEGRATED Q-ENHANCED RF FILTERS FOR MULTI-BAND/MODE WIRELESS APPLICATIONS

Gao Zhiqiang, Associate Professor

Department of microelectronics of Harbin Institute of Technology China

SECTION I: WIDEBAND RECONFIGURABLE CMOS GM-C FILTER FOR WIRELESS APPLICATIONS

Introduction

Recent developments in portable applications and systems have lead to a significant in wireless standards. Therefore, cost efficiency of CMOS technology implementation has been greatly enhanced with the emergence of multi-mode wireless applications. Now multimode/multi-band receivers are designed based on the scheme of reuse[1]-[3]. They avoid using multiple chipsets and can be made tunable which makes them more efficient in term of area and power consumptions. In a flexible receiver front-end, analog baseband filtering is a key task as it is used to select the required information under desired channel bandwidth. A large tuning range of the band-pass filter should be required for various wireless applications. To meet different specifications for the desired channel in multimode receivers, there has been a tremendous amount of research [1-6] effort aimed at improving the performance of integrated reconfigurable continuous-time (CT) filters in recent years. However, due to the open-loop operation nature, Gm-C filters generally use operational transconductance amplifier (OTA) driving a capacitor load at the cost of moderate linearity, sensitivity to parasitics. Moreover, the major disadvantage of OTA is the large distortion caused by the nonlinear behavior of the transistors involved. To enhance the linearity of the OTA and avoid potential stability problems, an approach to linear Gm-C integrators with inherent CMFB is developed based on the techniques of the cross-coupled differential pairs and source degeneration with passive resistors. In Section 2, the high linear transconductor Gm is presented. The design of the Chebyshev bandpass filter is discussed, as such the simulated results of the bandpass filter are given in Section 3. The conclusion is given in Section 4.

THE LINEARIZED TECHNIQUES OF TRANSCONDUCTORS

The transconductor in CMOS process is required for wideband reconfigurable Gm-C filter. Thus, the following section discusses the reported basic linearity technique in CMOS process. The transconductor linearity techniques can be broadly classified into three types: (a) source degeneration (b) cross coupling (c) active biasing.

Source Degeneration

Figure 1 shows circuit implementation of the source degeneration technology. The feedback equivalent resistance R (or M3, M4) is called source degeneration resistor, and differential pair M1, M2 and source degeneration resistors consist of the structure of source degeneration. The output current of the structure is related to the input voltage by the following equation

$$I_O = I_{D1} - I_{D2} = (V_d - I_O R)\sqrt{2KI_{SS}}\sqrt{1 - \frac{K(V_d - I_O R)^2}{2I_{SS}}}$$

(2.1)

Where $K = \frac{1}{2}\mu_0 C_m \frac{W}{L} \cdot I_{SS}$ is tail current source of the transconductor as shown in Figure 1, and the nonlinearity term of (2.2) is $V_d - I_O R$.

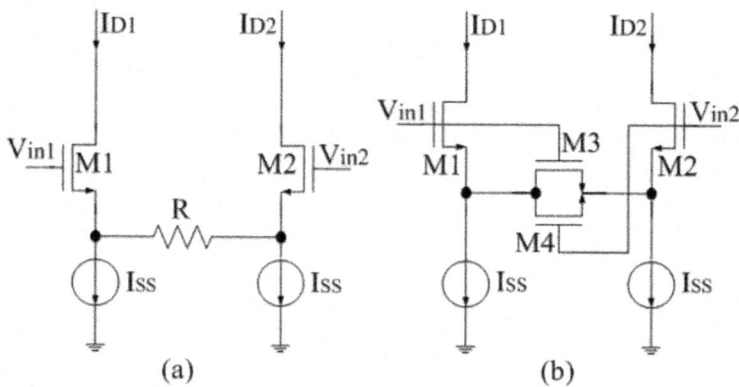

Figure 1: The typical source-degeneration structure of transconductor.

The transconductance Gm of source–degeneration structure can be about expressed as

$$Gm \approx \frac{g_m}{1+g_m R}$$

$$(2.2)$$

When the resistance is much greater than 1/gm, the transconductance $Gm \approx 1/R$. However, this is traded-off with the noise and power consumption. In CMOS process, high quality passive resistance is achieved difficultly.

Cross Coupling

A simple differential pair can cancel out the even order harmonics of distortion of transconductor output current. The remaining odd order harmonics can be cancelled out by two cross-coupling differential pairs with the same distortion but with different gm values. The circuit is shown in Figure 2.

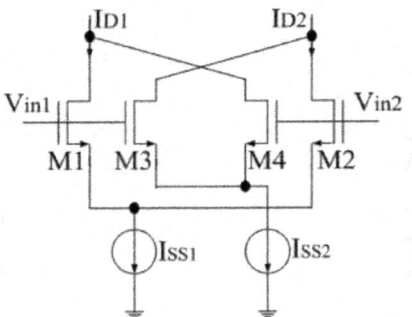

Figure 2: The transconductor with differential cross-coupled pairs.

The 3rd order harmonic is the main concern since it is now the most significant distortion. From Eq. (2-3), the 3rd order harmonic distortion (HD3) of output current can be obtained as:

$$HD_3 = \frac{K^{3/2}}{2\sqrt{2I_{SS}}} V_d^3$$

$$(2.3)$$

Since HD3 depends on the ratio of K3/2 and Iss1/2 only, the distortion can be cancelled by connecting two differential pairs M1, M2 in parallel with M3, M4 as shown in Figure 2. The transconductor parameter $K_{3,4}$ and $K_{1,2}$ are related to I_{SS2} and I_{SS1} as follows:

$$\left(\frac{K_{3,4}}{K_{1,2}}\right)^3 = \left(\frac{(W/L)_{3,4}}{(W/L)_{1,2}}\right)^3 = \frac{I_{SS2}}{I_{SS1}}$$

$$(2.4)$$

The corresponding effective gm is then given by:

$$g_{meff} = g_{m1,2}[1 - \left(\frac{K_{3,4}}{K_{1,2}}\right)^2] = g_{m1,2}[1 - \left(\frac{I_{SS2}}{I_{SS1}}\right)^{2/3}]$$

(2.5)

According to equation (2.5), when $I_{SS2} \ll I_{SS1}$, the transconductance is approximated as linearity. But the noise performance is worse than that of a simple differential pair because 2 differential pairs are connected. However, the noise is not doubled because $K_{3,4} < K_{1,2}$.

Active Biasing

The idea of active biasing is to make the biasing current compensate for the non-linear term:

$$I_{ss} = I_{DC} + \frac{KV_d^2}{2}$$

(2.6)

Where I_{DC} is the DC bias current, and V_d is input differential signal. Now the bias Iss supplies I_{DC} when $V_d = 0$ for the static bias. When there is a signal, an additional bias current $KV_d/2$ will compensate for the drop of the gm. This can be verified by inserting the new I_{ss} into Eq. (2.7):

$$I_{D1} + I_{D2} = 2K(V_{gs} - V_{th})^2$$

(2.7)

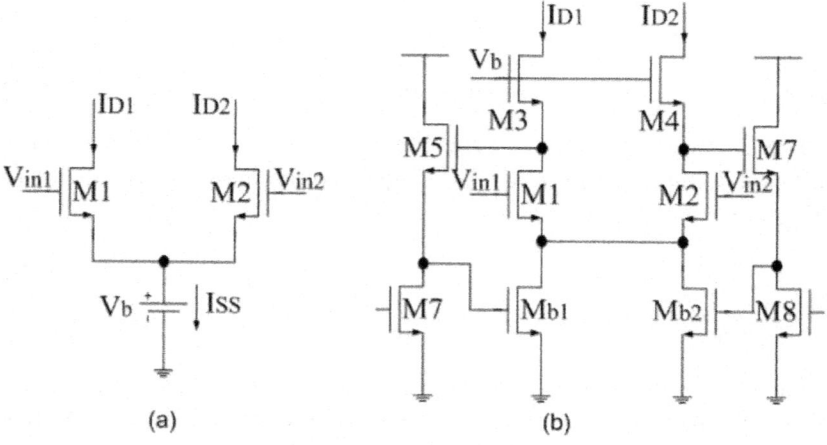

(a)   (b)

Figure 3: The transconductor with active-biasing differential pair.

In this design, all transistors are matched except M5-M8. For this cascode circuit, when there is an input signal, the same amplitude appears at the drains of M1 and M2 because the loading is $1/gm_{3,4}$ and $gm_{3,4} = gm_{1,2}$. The capacitance at that node and the loss of the level shifter M5-M8 are ignored. Both gates of Mb_1 and Mb_2 sense the differential voltage. Because the drains of Mb_1 and Mb_2 are connected together, a bias current is obtained as in Eqs. (2-26). The required active biasing is then established. But in the common-mode sense, now the conductance of Mb_1 and Mb_2 increases in phase with the input signal. This is a kind of feedforward and thus causes a boost-up of the common-mode gain. As a result, CMRR drops and common-mode instability will be resulted.

The Design of High Linear Transcoductor

The OTA is based on the Gilbert multiplier, which uses the two cross-coupled differential pairs (M1 - M2, M3-M4) as the input stage to reduce the nonlinearities as shown in Figure 6. Thank to mismatching of passive resistor in CMOS process, active source-degeneration resistor M_{R1} and M_{R2} is perfect choice. For this OTA structure, all transistors operate in the saturation region except for the transistors M_{R1} and M_{R2}. The MOS transistor is approximated as

$$\frac{1}{R_{eq}} \approx K_R(V_R - V_{th} - V_{CMS})$$

(2.8)

Where K_R is relative to MOS process parameter, $V_{R1,2}$ is control voltage of source degeneration resistor, and V_{CMS} is common-mode voltage of tail current source. The output current of the transconductance is

$$I_O = I_{O1} - I_{O2} = (I_{d1} - I_{d3}) - (I_{d2} - I_{d4}) = (I_{d1} + I_{d4}) - (I_{d2} + I_{d3})$$

$$= \sqrt{2KI_{DC1}}V_d\sqrt{1 - \left(\frac{V_d}{2V_{dsat1}}\right)^2} - \sqrt{2KI_{DC2}}V_d\sqrt{1 - \left(\frac{V_d}{2V_{dsat2}}\right)^2}$$

2.9)

Where V_d is input differential voltage, I_{DC1}, I_{DC2} is drain terminal current M_{b1}-M_{b4} respectively. V_{dsat1}, V_{dsat2} is source-drain overdrive voltage M_{b1}-M_{b4} respectively. Expanding (2.9) in the Taylor series and considering the first three terms only, the even terms for differential is neglected, and (2.10) becomes

$$I_O = I_{O1} - I_{O2} \approx (g_{m1} - g_{m2})V_d - \frac{1}{8}\left(\frac{g_{m1}}{V_{dsat1}^2} - \frac{g_{m2}}{V_{dsat2}^2}\right)V_d^3$$

(2.10)

Taking into account the mobility degradation, equation (2.10) is expressed as

$$I_O = I_{O1} - I_{O2} \approx (\frac{g_{m1}}{1 + g_{m1}R_1'} - \frac{g_{m2}}{1 + g_{m2}R_2'})V_d$$

$$-\frac{1}{8}(\frac{g_{m1}}{V_{dsat1}^2(1 + g_{m1}R_1')^3} - \frac{g_{m2}}{V_{dsat2}^2(1 + g_{m2}R_2')^3}V_d^3$$

(2.11)

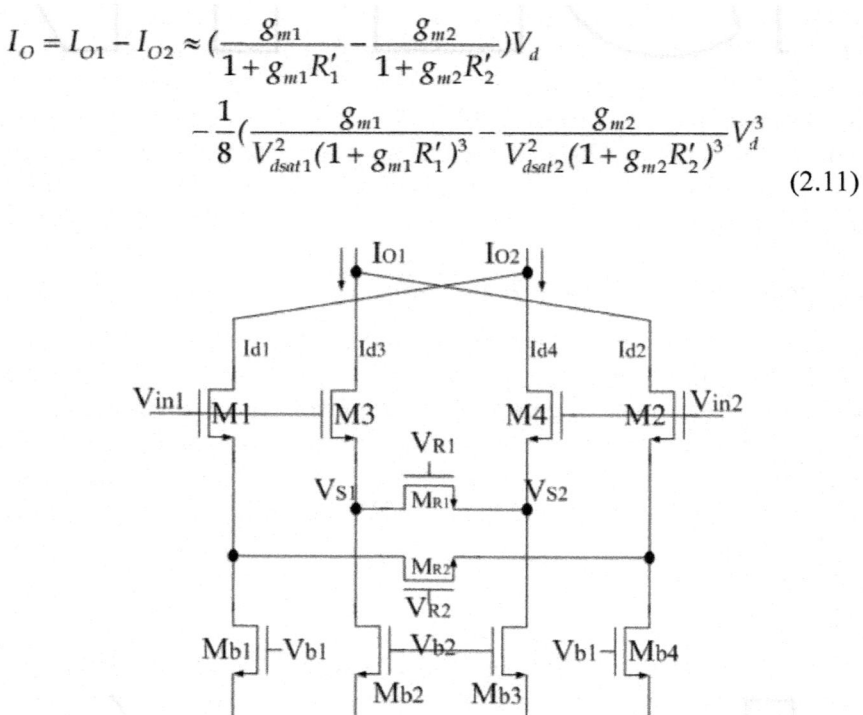

Figure 4: The Gilbert OTA with source degeneration.

Where $R_1' = R_1 + R_{\theta1}$, $R_2' = R_2 + R_{\theta2}$, $R_{\theta1}$, $R_{\theta2}$ are source series resistance of the mobility degradation, $R_\theta = \frac{2\theta}{K}$, and θ is the mobility reduction coefficient.

In equation (2.11), if the nonlinearity third-order term satisfies

$$\frac{g_{m1}}{V_{dsat1}^2(1 + g_{m1}R_1')^3} - \frac{g_{m2}}{V_{dsat2}^2(1 + g_{m2}R_2')^3} = 0$$

(2.12)

Then the transconductance is expressed as

$$Gm = \frac{g_{m1}}{1 + g_{m1}R_1'} - \frac{g_{m2}}{1 + g_{m2}R_2'}$$

(2.13)

If the condition $R \gg 1/g_m$ is satisfied, the transconductance can be obtained by

$$G_m \approx \frac{1}{R_1'} - \frac{1}{R_2'}$$

(2.14)

Figure 5 is overall structure of the high linear transconductor. Figure 6 shows the simulation of step response of the transconductance. When the dc common-mode voltage is about 1.67V, the transient-time response is less than 60ns, and the variation of common-mode voltage is less than 15mV. We use 5pF as the loading capacitance to verify the AC response of the transconductor. The bandwidth of unit gain is about 98MHz, and the phase margin is about 76 degree as shown in Figure 7.

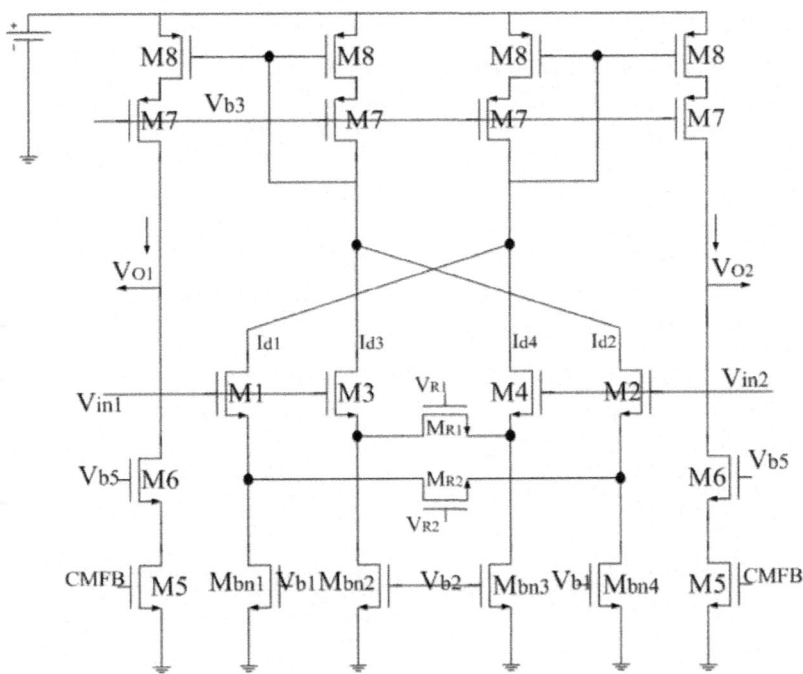

Figure 5: The overall structure of high linear transconductor.

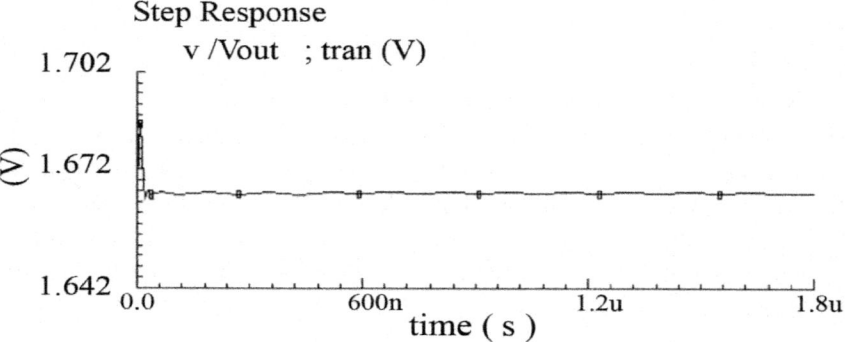

Figure 6: Step response of the proposed transconductor.

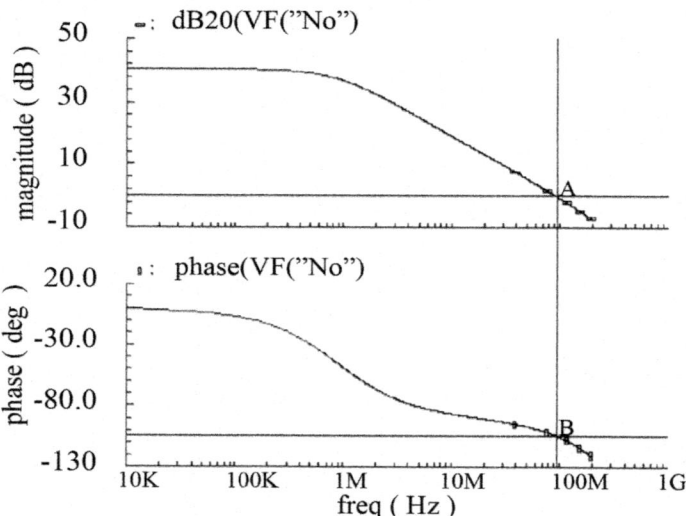

Figure 7: The response of amplitude and phase for the transconductor.

Figure 8 shows the simulated Gm plot with the input voltage. The linear range of the proposed active degenerative resistance (ADR) Gm of cross-couple differential pair is about ±1V. The linear range is higher than the other Gm of differential cross-coupled pair without ADR and differential pair with ADR (source degeneration structure as shown in Figure 2). When the operating frequency is 4MHz, the third-order intermodulation IM3 is -72dB as shown in Figure 9.

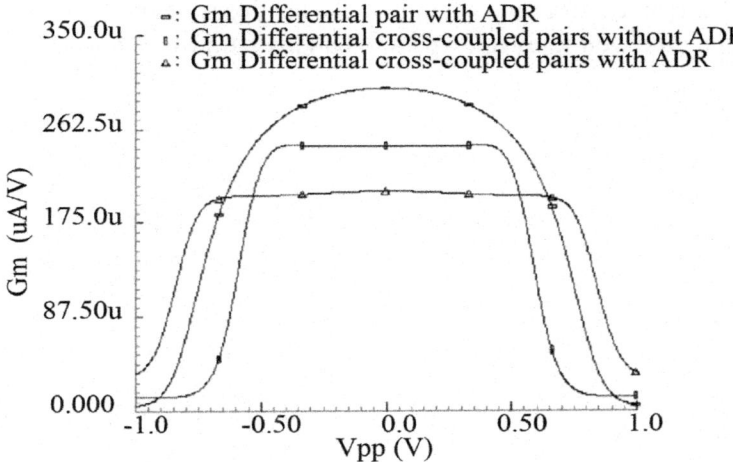

Figure 8: Simulation of linearity for the three differential linearized transconductor.

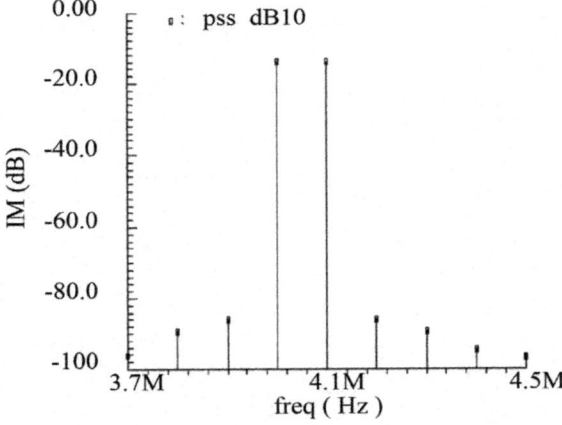

Figure 9: Response of third-order intermodulation distortion of the OTA at 4MHz.

Circuit Design of Gm-C Filter

Using the proposed high linear OTA, a six-order Chebyshev bandpass filter is designed. To obtain a passband frequency with minimal sensitivites to individual component values, the filter topology is derived from a doubly terminated passive RLC lowpass prototype as shown in Figure 10. The associated signal flow graph (SFG), shown in Figure 11 is obtained by writing down the state equations for six reactive components. Notice that the SFG contains only integrators and summers. The bandpass filter topology is derived by applying the well-known lowpass to bandpass transformation

$$S_L = \frac{s^2 + \omega_0^2}{s\omega_C}$$

(2.15)

Where S_L is the normalized lowpass Laplace variable, ω_0 is the center frequency, and ω_C is the bandwidth of the bandpass filter to be designed.

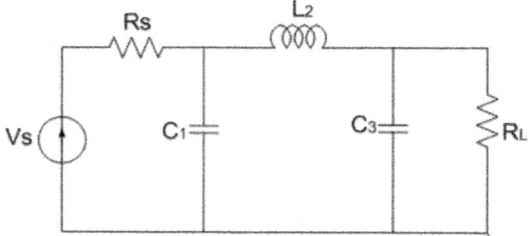

Figure 10: The passive RLC lowpass prototype of low-pass filter.

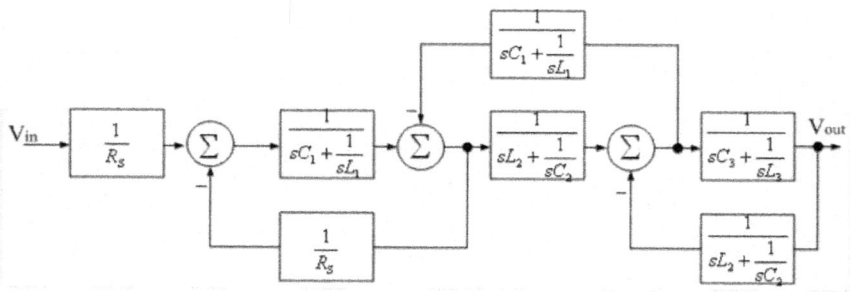

Figure 11: The signal flow of the transfer function from passive low-pass filter to bandpass filter.

The design method is based on component substitution. A ground inductor produces two transconductors and one capacitor, while a floating inductor need four transconductors and one capacitor as shown in Figure 12(a),(b). Figure 12(c) shows the use of a differential transconductor connected as a pseudo-resistor. The bandpass filter topology is obtained by replacing six integrators with six coupled resonators.

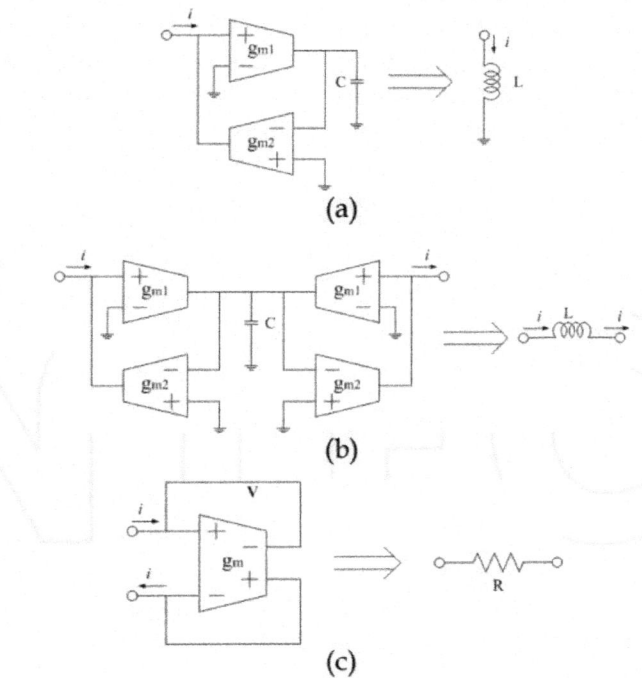

Figure 12: Methods of passice component substitution.

The complete filter is shown in Figure 13. The resonator is composed of two Gm-C integrators with feedback loop.

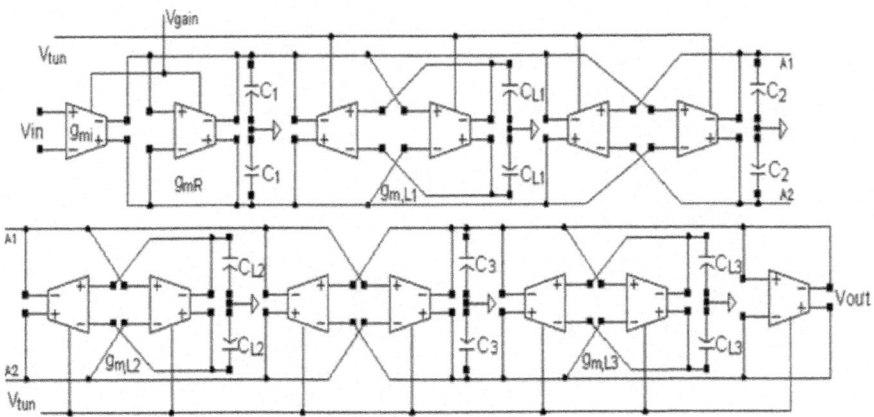

Figure 13: The sixth-order Chebshev OTA-C bandpass filter.

The proposed filter is simulated with Cadence's Spectre softwares using TSMC 0.25um standard CMOS process models. Simulation results in Figure 14 and Figure 15. Figure 13 shows the AC tuning-Q response of the filter when the tuning center frequency is about 2MHz. The center frequency tuning range is about 0.5MHz to 10MHz as shown in Figure 14.

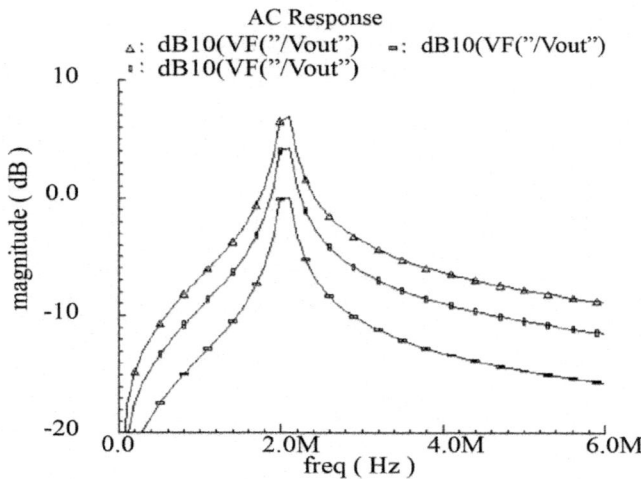

Figure 14: The tuning-Q response of the Gm-C bandpass filter at fc≈2MHz.

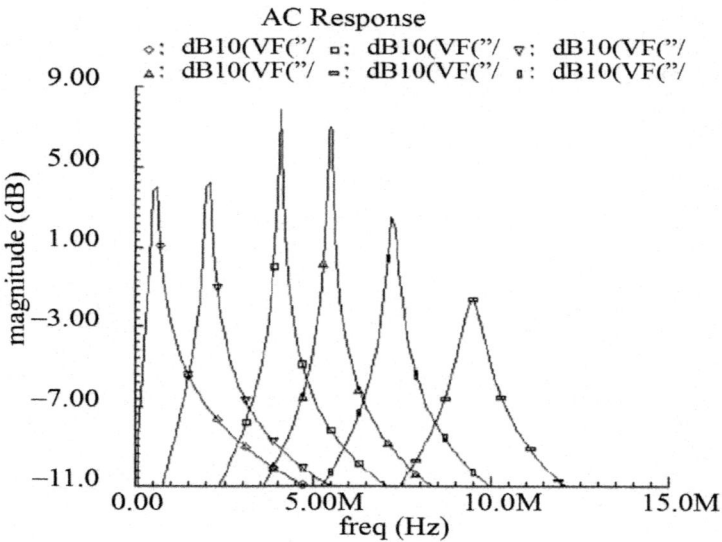

Figure 15: Tuning center frequency of the OTA-C bandpass filter.

CONCLUSION

A full CMOS six-order Gm-C Chebyshev filter based on passive LC-ladder synthesis is designed in TSMC standard 0.25um CMOS process, which uses a highly linear operational transconductance amplifier (OTA) based on cross-coupled differential pairs with sourcedegeneration structure, which exhibits a wide product of gain-bandwidth, and high linearity. The simulated results show the center frequency tuning range of the filter is from about 0.5MHz to 10MHz and the maximum quality factor of 150 at the center frequency 4.3MHz. The filter is suitable for multi-band wireless applications.

SECTION II: A RF LC Q-ENHANCED CMOS FILTER FOR WIRELESS RECEIVERS

INTRODUCTION

Despite decades of research in developing "single-chip" radio transceivers, most designs continue to rely on off-chip components for RF bandpass filtering. Implementing these filters on-chip remains nearly as challenging today due to problems in meeting system requirements. Recent advances in silicon-on-chip IC processes targeted at RF designs, however, offer the possibility of producing on-chip filters in the coming years using Qenhancement techniques. CMOS technology is an attractive solution due to the low cost, high-level integration. One of prevalent off-chip component required in wireless receiver circuits is RF bandpass filter, usually realized with a surface acoustic wave (SAW) or ceramic device. If on-chip high frequency filters with acceptable electrical characteristics can be realized, this would eliminate or reduce the need for these currently required off-chip filters. This implementation of integrated filters could lead to complete communications system design solutions on monolithic chip that would decrease the complexity, reduce the size, lower the power and cost of wireless transceiver circuits. However, the use of on-chip RF bandpass filters in commercial radio transceivers has been limited so far by inferior performance relative to system requirements. Such requirements include: narrow bandwidths, high linearity, low insertion and noise figure, and the need for low-power consumption in wireless system of aerospace applications. This fact has made the acceptance of on-chip designs much more difficult than it would be if the system specifications were more relaxed, pushing radio designers to embrace modified, and generally problematic, radio architectures such as the direct-conversion, or zero-IF schemes. If on-chip bandpass filters are accepted into commercial products, they must at least compete with the performance of products designed around these architectures. In this section, we outline the alternatives for building on-chip bandpass filters practical considerations in section 2. The design of the integrated on-chip 1.8V 2.14GHz Q-enhanced LC filter using silicon CMOS process is presented in section 3. The simulated results of the filter presented are provided in section 4. Finally in section 5, conclusion is drawn.

Filter Technology

A wide range of technologies exists for implementing RF bandpass filters, as illustrated in Figure 16.

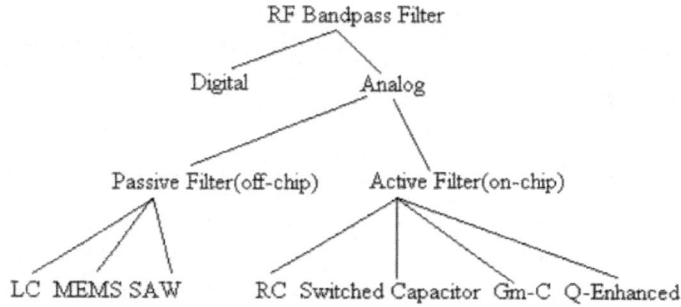

Figure 16: Taxonomy of RF bandpass filter implementations.

In theory, digital filters could implement any filter desired, and achieve true "software-radio" realizations. However, their applications are still generally limited to baseband frequency due to problems with power consumption and noise at high frequencies. To illustrate these problems, bounds on power consumption were developed in based on CV^2f considerations, because the scale of the digital filter has at least thousands of transistors and each of the transistors is considered as source of the noise. If the digital filter is operated at RF band, the power consumption and noise of the filter is large enough to beyond imagination. Currently, however, on-chip RF bandpass filtering remains an analog endeavor. Here, the choices involve passive or active designs, each with several implementation possibilities including LC, and fully active architectures. In these active architectures, RC filter and switched capacitor filter isn't suited for high frequencies application. Although Gm-C filter can be operated at high frequencies, it poses its own drawbacks, namely, large power consumption and high noise contribution. A promising solution is to implement a active LC filter using on-chip passive elements with loss compensation circuitry to improve the effective quality factor [1, 3-7, 10].

Q-Enhanced Filter Design

On-chip spiral inductor

The design and characterization of on-chip inductors is central to the implementation of high performance Q-enhanced LC filters. In a typical two-metal silicon IC process, these inductors are fabricated using planar spiral geometries as illustrated in Fig. 17. Top layer metallization usually provides the lowest resistivity and capacitance to the underlying substrate and is therefore used for the spiral turns, while the lower metallization layer is used for connection to the spiral center.

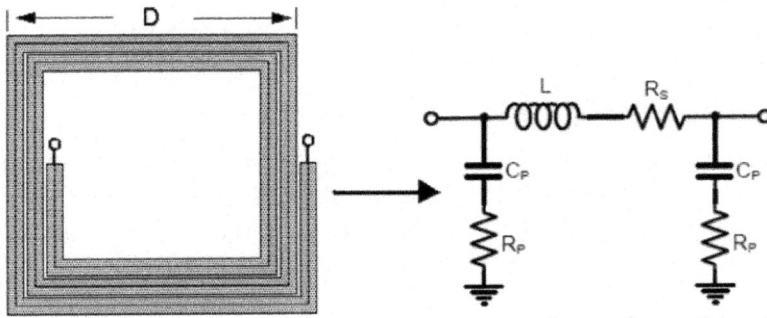

Figure 17: Top view of an on-chip spiral inductor and its electrical model.

In practice, electrical characteristics of the integrated inductor are generally frequency dependent and are more precisely described with a lumped-element model of greater complexity. A commonly used square-spiral IC layout and a more accurate electrical model for the inductor, the π-model, are shown in Figure 2. In this model, Rs represents the resistive losses in the metal traces of the inductor, any contact losses, and losses attributable to eddy currents in the substrate. Note that the top branch of Figure 2, which is composed of series resistor RS, along with the inductance L represents the simplified series resistance model for the inductor shown in Figure 2. The substrate capacitance is modeled by C_p, and R_p represents loss caused by substrate conductance. With the help of a patterned ground shield, the electric field from the lossy substrate [3][10]. Thus, the only dominant loss due to the serious resistance RS, and to cancel it, it is necessary to implement a loss compensation mechanism that effectively introduces a negative resistance of the same magnitude in series with R_S.

Q enhancement

A primary method for increasing the Q of non-ideal on-chip resonators is through the use of active devices to create negative resistance. Although methods that include phase-shifted current feedback via coupled inductors [21] have been investigated, the direct use of active devices as negative resistors is the prevalent Q enhancement technique. Single-ended negative resistance methods have been documented [23-25], while the more common differential method using a cross-coupled transistor pair is presented in Figure 18. The voltage to current ratio indicates the effective negative resistance at the terminals of the cross-coupled MOSFET shown in the figure and is described by

$$R = -\frac{2}{g_{mQ}}$$

(2.16)

It is clear from Figure 18 and Equation (2.16) that the effective negative resistance can be adjusted by changing the bias source, I_Q, and thereby the transconductance, g_{mQ}, of the differential pair MQ1, MQ2. This facilitates electronic tuning of this loss-canceling mechanism.

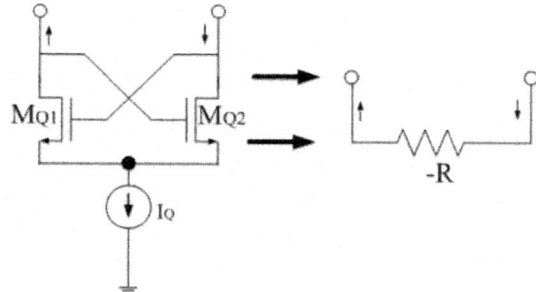

Figure 18: Cross-coupled MOSFET negative transconductance.

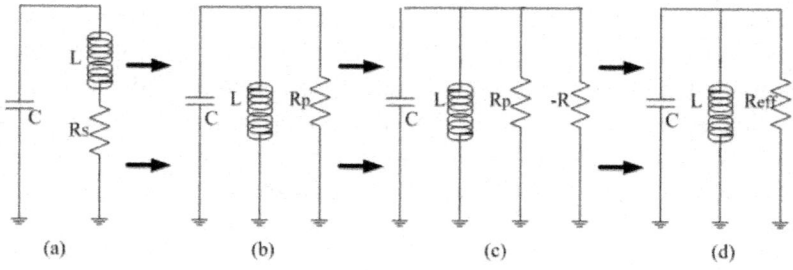

Figure 19: Q-enhanced LC circuit.

The concept of Q-enhancement for an LC tank circuit with parallel-connected negative resistance is illustrated in Figure 18-19, with the series resistance inductor model utilized to simplify the analysis. We assume for now that a lossy inductor and a capacitor can be simplistically modeled as shown in Figure 19(a), in which R_S represent the losses in the inductor. We can compensate the losses by connecting negative resistor in parallel with the LC tank as shown in Figure 19(c). In this approach, the negative resistance has been implemented negative resistance –R as shown in Figure 17 to cancel the loss represented by R_p. The effective parallel resistance R_{eff} and the effective quality factor Q_{enh} of the LC resonator as shown in Figure 19(d) is given by

$$R_{eff} = R_p // \frac{-1}{g_m} = \frac{1}{1 - g_m R_p} R_p$$

(2.17)

and

$$Q_{enh} = \frac{R_{eq}}{X_L} = \frac{1}{1 - g_m R_p} Q_0$$

(2.18)

Where Q_0 is the self-tuning quality factor of the circuit as illustrated in Figure 19(a). It should be noted that as gmRp continues to increase and approaches the unit, the value of Q_{enh} is infinite and the circuit (theoretically and practically) become an oscillator. The prototype circuit of the second-order RF Q-enhanced bandpass filter based on the above negative-resistance techniques is shown in figure 20. Common-source transistor M_1 and M_2 are employed for the input buffer stage. They are large devices and biased to have low-input impedance and small distortion. Two pair of PMOS transistors M_C are varactors which are used to tune the center frequency of the filter and also adjust the quality factor of the filter through DC voltage V_{con}. The PMOS M_C capacitors are operated in depletion and inversion regions within the tuning range. The negative-resistance circuit can be realized by crosscoupled differential pair M_{Q1} and M_{Q2}. The negative resistance is $-2/g_m$ if we assume that the two transistors have the same size. As such, the negative resistance can be adjusted by current source IQ.

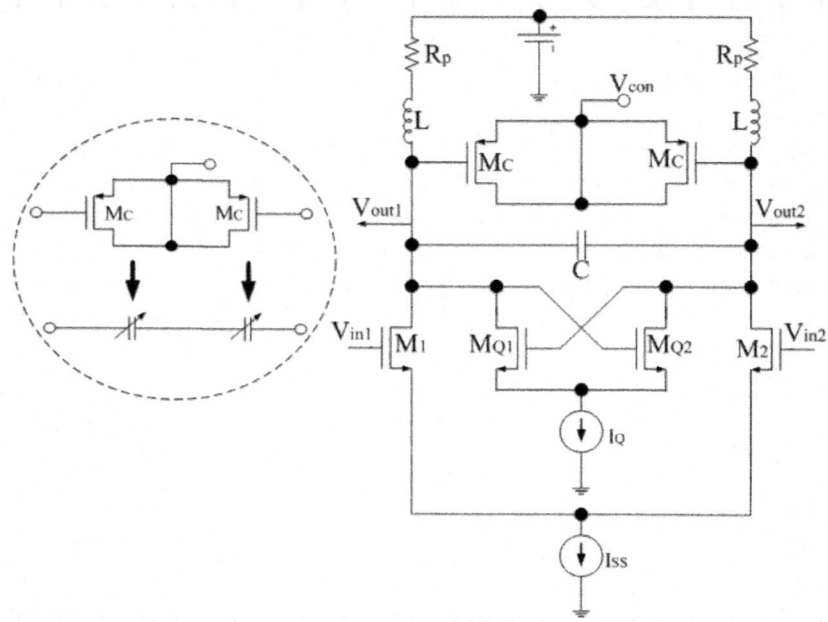

Figure 20: A 2nd Q-enhanced on-chip LC bandpass filter.

The transfer function, quality factor Q, and center frequency ω_0 of the biquad Q-enhanced filter can be expressed as

$$H(s) = \frac{\dfrac{g_{m,in}}{C_{tot}}(s + \dfrac{R_p}{L})}{s^2 + (\dfrac{R_p}{L} - \dfrac{g_{mQ}}{C_{tot}})s + \dfrac{1}{LC_{tot}}(1 - g_{mQ}R_p)}$$

(2.19)

$$Q = \frac{R_p C_{tot}}{\sqrt{LC_{tot}}(1 - g_{mQ}R_p)}$$

(2.20)

$$\omega_0 = \frac{1}{\sqrt{LC_{tot}}}$$

(2.21)

Where $C_{tot} = C + 2C_{gmc} + 2C_{gm,in} + 2C_{gmQ}$, C_{gmc} is the parasitic capacitance of transistor Mc, $C_{gm,in}$ is the parasitic capacitance of input buffer M_1, M_2, C_{gmQ} is the parasitic capacitance of transistor M_{Q1}, M_{Q2}. $g_{m,in}$ and gmQ are the transconductance values of the input buffer M_1, M_2 and Q-enhanced cross-coupled differential pair M_{Q1} and M_{Q2}, respectively. To facilitate a theoretical noise analysis of the circuit, an appropriate noise model for the MOSFET and passive components is required. For noise modeling, passive L and C components will be considered noiseless. The accepted expression is used for passive resistors and transistors. As shown in Figure 19, the mean square noise contributions of each component at the center frequency are given by

$$NF = 1 + \frac{8kT\gamma(g_{mQ} + g_{m\Phi in}) + 8kTR_p}{4kTR_s g_{m,in\Phi}^2}$$

$$(2.22)$$

Simulation Results

To verify the design of the Q-enhanced RF bandpass filter, the filter was simulated with TSMC 0.18μm CMOS technology. At the same time, in order to test the circuit, the external wideband transformers are employed to serve as the impedance matching. The quality factor Q, gain S21, and noise performance is simulated with Cadence Spectre tools as shown in Figure 21-23. The input reflection coefficient S_{11} is about –23dB when the center frequency is about 2.14GHz and bandwidth is about 36MHz in Figure 24. The effect of Qtuning controls on the filter's response is shown in Figure 21 when the bias current IQ and I_{ss} is from 200uA to 500uA. The maximal value of the quality factor for the filter can be attained 60. Noise figure is about 15dB at center frequency of 2.14GHz in Figure 22. Figure 23 shows the gain of the filter at center frequency of 2.14GHz and Q=60, S_{21} is about 15dB. The linearity performance of the filter for f_c=2.14GHz and input power -60dBm is tested by IIP3 as shown in Figure 25. Two-tone signal at 2.14 and 2.144GHz is presented at the filter input through an RF power combiner, the input power at the filter input is -30dBm, which is small (the amplitude of the input voltage is about equals to 7mV) when the input load is 50Ohm. The third-order intercept point(IP3) is about -7.63dBm with SpectreRF PSS tools. The input noise floor is also measured by Cadence tools and the RMS value N_{out} is measured to be about -90.5dBm. Thus, the spurious-free dynamic range (SFDR)[2] can be calculated as 2

$$SFDR = \frac{2}{3}(IIP3 - N_{out}) \approx 56dB$$

$$(2.23)$$

The relation between the dynamic range and the quality factor of the filter is simulated as shown in Figure 26 for the center frequency of 2.14GHz. The simulation in Figure 25 shows that the dynamic range is lower when the Q is increased. The main reason for the degradation in the dynamic range is quality factor and the output noise voltage is increased, it leads to deteriorating the linearity of the filter. The performance of the filter is summarized in Table 1.

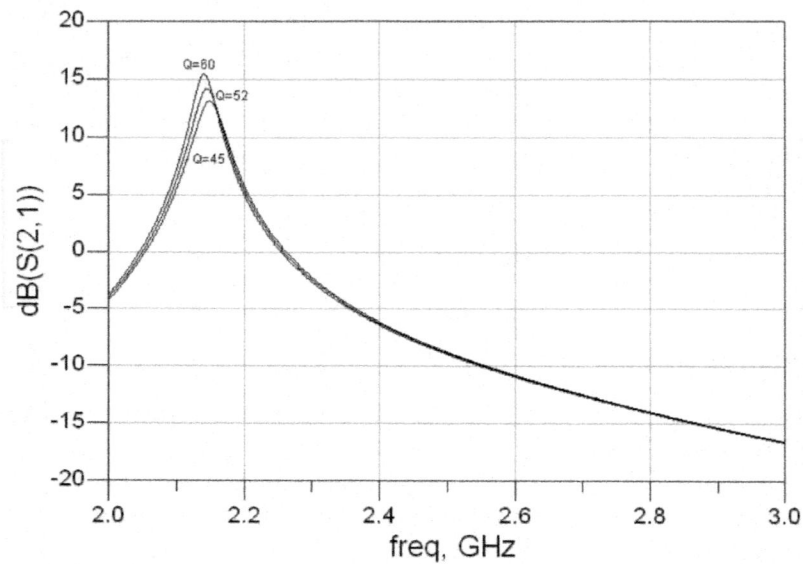

Figure 21: Quality factor tuning of the RF LC Q-enhanced filter.

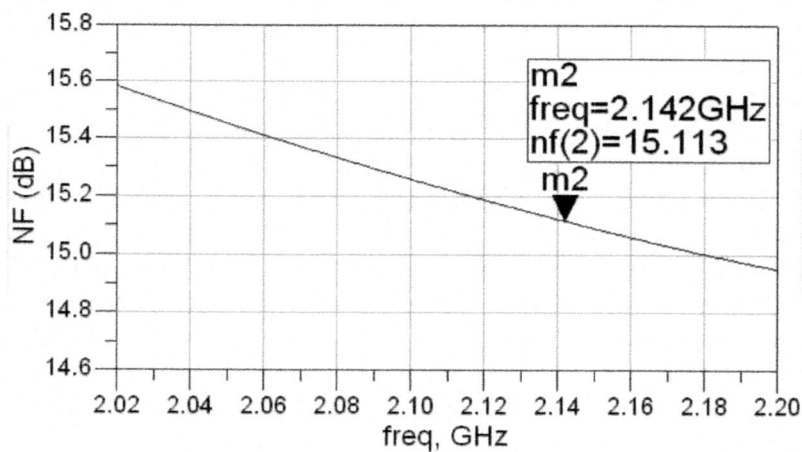

Figure 22: Noise figure of the RF LC Q-enhanced filter.

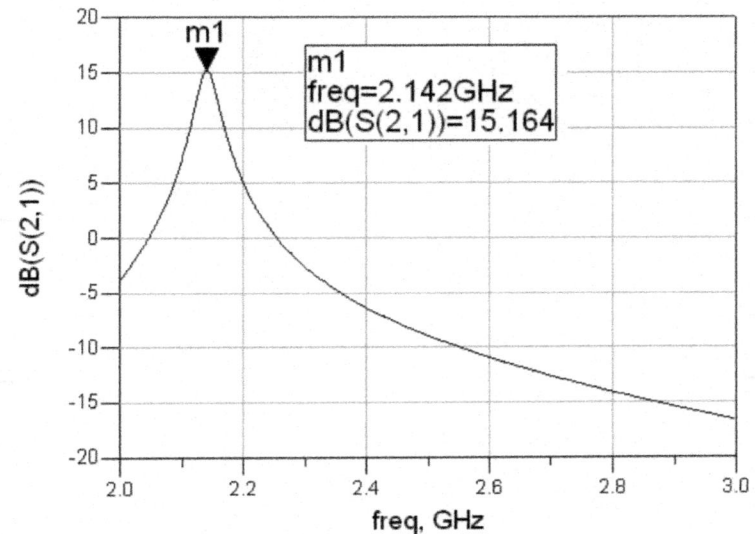

Figure 23: S21 response of the RF LC Q-enhanced filter.

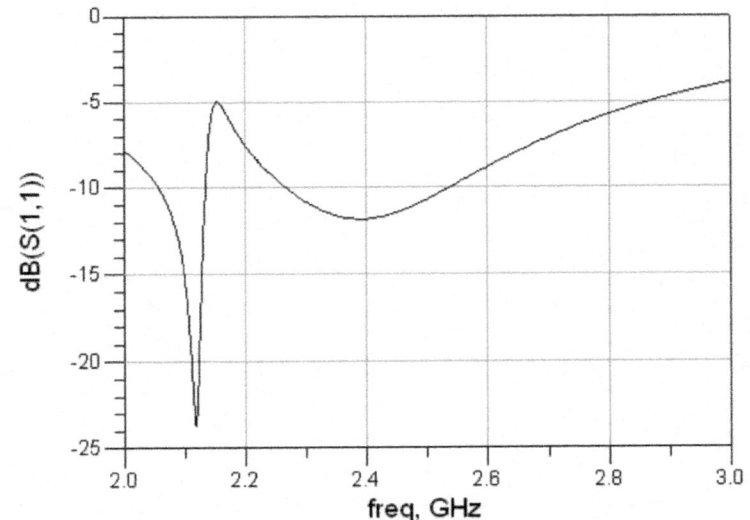

Figure 24: Input reflected loss response S11 of the RF filter.

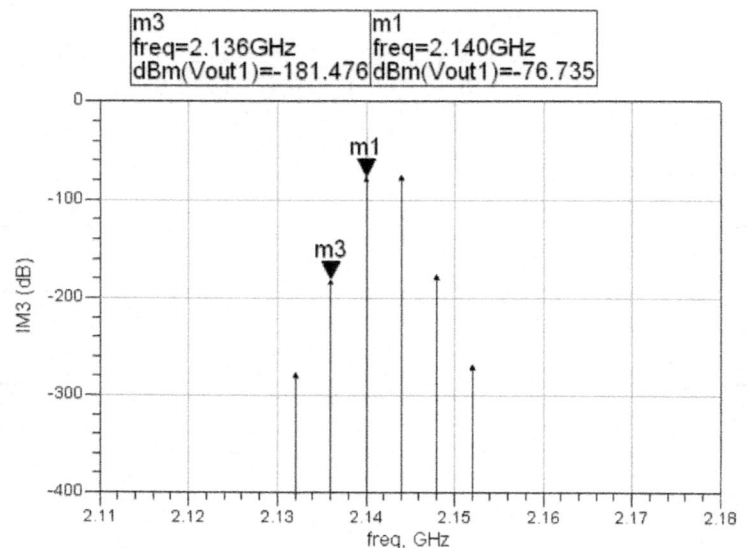

Figure 25: The 3rd order intercept point response of the RF filter.

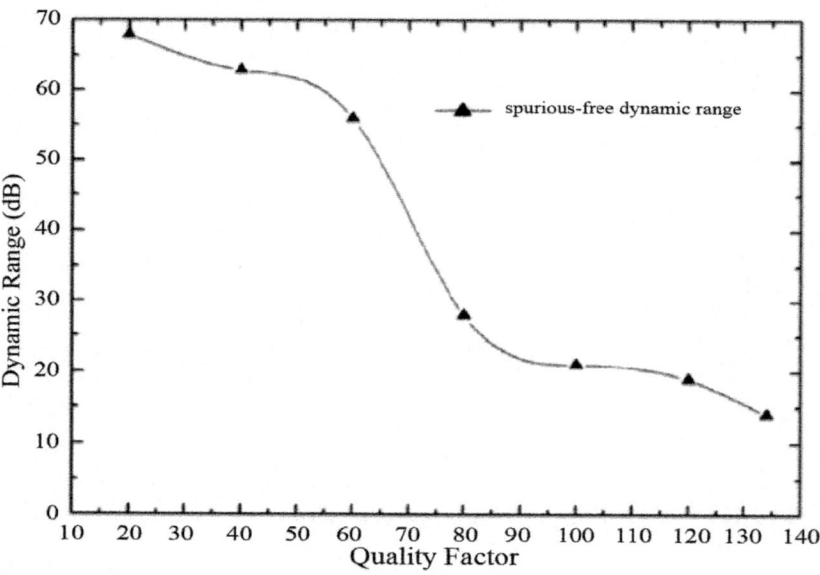

Figure 26: The related curve of the dynamic range and quality factor.

Table 1: The performance comparision of RF active integrated LC filter

Performance Parameters	[11]	[12]	This work
Technology	0.25µm CMOS	0.5-Si-SOI	0.18µm CMOS
Center frequency	2.14GHz	2.5GHz	2.142GHz
3-dB Bandwidth	60MHz	70MHz	36MHz
Maximum Gain in passband	0dB	14dB	15dB
Noise Figure	19dB	6dB	15dB
Supply voltage	2.5V	3V	1.8V
DC consumption	17.5mW	15mW	15mW

Table 1 shows the comparison for published CMOS, and bipolar RF integrated bandpass filters in the literature. The comparison table demonstrates that the proposed RF filter has lower power-supply, the highest selectivity, and the largest gain.

Conclusion

A 2.14GHz CMOS fully integrated second-order Q-enhanced LC bandpass filter with tunable center frequency is presented. The filter uses a resonator built with spiral inductors and inversion-mode MOS capacitors which provide frequency tuning. The simulated results are shown that the filtering Q and gain can be attained 60 and at 2.14GHz, and the spuriousfree dynamic range (SFDR) is about 56dB with Q=60 and power consumption is about 15mW. The presented filter is suitable for S-band wireless applications.

SECTION III: A FULLY INTEGRATED CMOS ACTIVE BANDPASS FILTER FOR MULTI-BAND RF FRONT-ENDS

Introduction

Now, the fast-growing market in wireless communications has led to the development of multi-standard mobile -terminals [1-3]. This creates a strong interest toward the highly integrated RF transceivers in a compact and low-cost way. So, it is becoming more and more attractive to have a single chip of the complete CMOS multi-band transceiver in the industrial, scientific, and medical (ISM) bands. However, the integrated high-performance filters working at RF frequency still remain the one of the most difficult parts in the integrated RF front-ends. The existence of large interference, spurious tones, unwanted image and carrier frequencies, as well as their harmonics in the wireless communication environment demands the use of RF filters with high selectivity in the RF front-ends as shown in Figure 27.

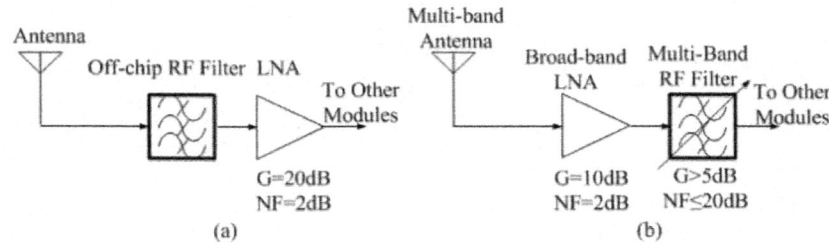

Figure 27: Multi-Band RF front-end designs.

In fact, in current gigahertz-range transceivers, the bulky and expensive off-chip bandpass filters [2] are still required to handle the existence of large out-of-band interference as shown in Figure 1(a). Furthermore, it increases the size, power consumption, and cost of multistandard transceivers significantly by adding different copies of discrete filters for different bands. Great efforts have been made to use an on-chip tunable Q-enhanced filter to replace such off-chip preselect filter. To this extent, recent researches on integrated filter design have fallen into the active-LC category [5]-[11]. Filters of this category are built around on–chip spiral inductors and capacitors used as LC resonant tanks, whereas an important cause for the limited integration of RF filters is the low quality factor of monolithic spiral inductors. These inductors are inherently lossy due to ohmic losses in the metal traces and due to substrate resistance and eddy currents. This problem has been addressed by using various methods such as patterned ground shields and geometry improvements, but the Q factor of integrated inductors is still generally limited to a value less than 20 [12] in standard RF CMOS process.

For multi-band RF front-end designs, a suitable on-chip tunable filter is available, but the tunable nature of the on-chip passive inductors is hard. Compared with the passive inductors, the RF bandpass filter using active inductors can not only achieve wide frequency tuning range and high quality factor, but also occupy the small chip areas. However, it also pays for the higher noise and the worse linearity. In commercial designs as shown in Fig. 1 (a), an LNA combined with a 3dB insertion loss discrete filter typically achieves a net 5dB noise figure, 17dB gain, and 1dB input compression point about -17dBm if the input P1dB of LNA is about -20dBm, while consuming 15mW [4]. If the filter using active inductors is located in the RF front-end as shown in Fig. 1(b), and the input P1dB of LNA is about 20dBm, the proposed RF filter and the LNA can achieve a net less than 4dB, and a net more than or equal to -20dBm input compression point with 15dB gain, so the proposed RF filter combined with other RF modules will satisfy the performance of the moderate noise figure and linearity of RF system requirements such as Bluetooth, 802.11b and so on.

The section is organized as follows. Section 2 presents the novel Q-enhanced active inductor topology, as well as the analysis of the noise figure linearity and stability. Section 3 describes the RF bandpass filter based on the active inductors and the measured results of the filter are demonstrated. Finally, conclusion is given in section 4.

Circuit Principle

Proposed Active Inductor

An often–used way for making active inductors is through the combination of a gyrator and capacitor, but designing high-Q active inductors at GHz with opamps or standard transconductance-C techniques is very difficult due to relatively significant power consumption and noise. The active inductor based on the principle of gyration, consisting of minimum-count transistors can be operated at GHz easily because fT of single transistor is so high as hundreds of GHz. A class of active inductors have been proposed by researchers [14] [19][20] in Figure 27. A common feature of these active inductor topologies is that they all employ some kind of shunt feedback to emulate the inductive impedance in Figure 28.

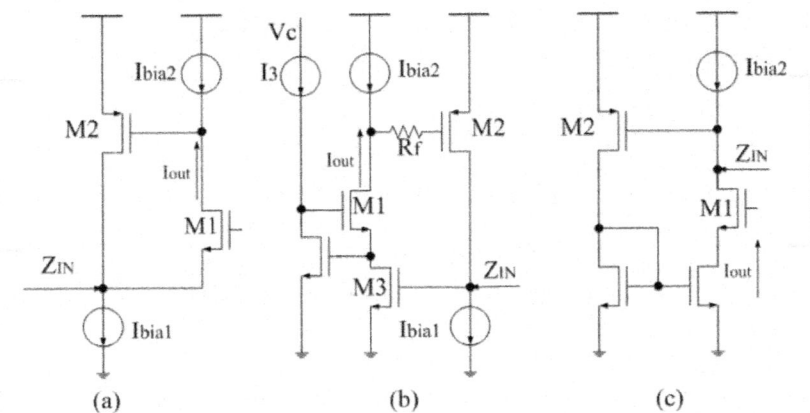

Figure 28: The proposed CMOS active inductor topology.

Intuitively, the circuits can be explained as follows: the input signal at the source of M2 will generate a current $g_{m2}V_i$ at the drain of M2, this current will be integrated on the gate-source capacitance Cgs1. The voltage at the gate of M1 will then generate the input current, thus generating the inductive loading effect. Compared with the active inductor proposed in Figure 28(a) and improved (b) or (c), we found the active inductor in Fig. 28(a) has some advantages over the active inductor in Figure 28(b) or (c). As can be seen from the circuit figure, the minimum voltage for the active inductor itself is only $\max(V_{gs1}+V_{ds1}+V_{in},\ V_{gs2}+V_{ds2}+V_{gs1}+V_{in})$. Therefore, the circuit in Fig 28(a) is better than the circuit in (b) or (c), and it has two transistors contributing noise directly to the input. In our design, the currentreused active inductor based on (a) is chosen.

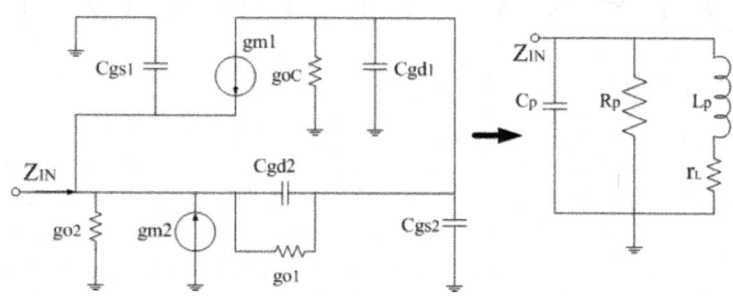

Figure 29: The small-signal equivalent circuit of the proposed active inductor

A conceptual illustration of the proposed active inductor is shown in Figure 27. A more detailed small- signal representation of Figure 28(a) is shown in Figure 29, where g_o is the drain-source conductance and g_{oc} represents the loading effect of the nonideal biasing current source Z_{load}. The impedance of Z_{in} can be expressed as

$$Z_{in} \equiv \frac{V_{in}}{I_{in}} = R_p\,//\,C_p\,//\,Z_L$$

$$(2.24)$$

where the inductive impedance of Z_{in} is

$$Z_L = \frac{g_{oc}+g_{o1}+s(C_{gs2}+C_{gd2}+C_{gd1})}{g_{m1}g_{m2}+[g_{m2}-g_{m1}+g_{oc}+s(C_{gs2}+C_{gd1})](g_{o2}+sC_{gd2})}$$

$$(2.25)$$

The small-signal analysis of the circuit in Figure 28(a) shows that Zin is a parallel RLC resonant tank with the following values:

$$Rp = \frac{1}{g_{o2}}\,||\,\frac{1}{g_{m1}} \approx \frac{1}{g_{m1}} \quad C_p = C_{gs1} \quad L_p \approx \frac{C_{gs2}}{g_{m1}g_{m2}} \quad r_L \approx \frac{g_{oc}+g_{o1}}{g_{m1}g_{m2}} \quad (2.26)$$

where r_L is the intrinsic resistor of the active inductor. The self-resonant frequency $\omega0$ and intrinsic quality-factor of the inductor is

$$\omega_0 \approx \sqrt{\frac{g_{m1}g_{m2}}{C_{gs1}C_{gs2}}} = \sqrt{\omega_{t1}\omega_{t2}}$$

(2.27)

and

$$Q_0 = \frac{R_p}{\omega_0 L_p} \approx \sqrt{\frac{g_{m2}C_{gs1}}{g_{m1}C_{gs2}}}$$

(2.28)

where ω_{t1} and ω_{t2} are the unity-gain frequency of M1 and M2, respectively.

Noise Analysis

Unlike the passive inductor where the damping resistor rL is the main noise contributor, the noise in active inductor originates from the thermal noise of MOS transistor channel [14], [15]. By referring to the transistor noise sources to the terminals of the active inductor in Figure 28(a), the noise figure of the circuit will be computed considering, for simplicity, only three main noise sources, i.e., the thermal noise of the two transistors (M1 and M2) and the noise of the load impedance Rp (i.e. $\frac{1}{g_o} || Z_{load}$). where $\overline{v_{n1}^2} = 4kT\gamma\Delta f / g_{m1}$, and $\overline{i_{n2}^2} = 4kT\gamma\Delta f g_{m2}$, kT is Boltzmann's constant times temperature in Kelvin, and γ is chosen empirically to match the observed thermal noise behavior of a given fabrication process. Computing the transfer functions from all noise sources to the output node, the following expression for the NF (at the resonance frequency) can be obtained

$$NF = 1 + \frac{\gamma}{g_{m1}R_S} + \gamma g_{m2}R_S + \frac{(1+g_{m1}R_S)^2}{g_{m1}^2 R_S \cdot R_P}$$

(2.29)

Where R_S is the source impedance. The second term in the right-hand side of (6) represents the noise contributed by transistor M1 and it has the same expression as for a common-gate amplifier. However, in this case, due to the feedback in the gyrator, g_{m1} can be made larger than $1/R_S$ while still ensuring matching conditions. The third term represents the noise introduced by the feedback transistor M2. Consistently with the intuition, transistor M2 injects noise directly at the input, and its transconductance has to be small to have a low noise. The fourth term in the equation represents the noise contributed by the load. If $g_{m1}R_S \gg 1$, this term becomes approximately equal to R_S/R_P. Notice that increasing R_P (i.e., increasing the quality factor of the resonant load) reduces the noise contributed by the load but also the noise of M2, since it results in a reduction of g_{m2}.

Nonlinear Distortion

As shown in Fig. 27, the distortion is mainly influenced by two factors: the additional current path provided by M2 and the effect of negative feedback on both the gate-source voltage swing across M1 and its DC bias point. The analytical expression for the circuit input P1dB can be found from Sansen's theory [13]. Considering the transistor in strong inversion, the input P1dB for the circuit as a function of the transconductance of transistors becomes

$$V_{in,1dB} = 2\sqrt{\left|\frac{0.244V_{in}^2}{1 - 2g_{m1}g_{m2}R_S R_p}\right|} \cdot (1 + g_{m1}g_{m2}R_S R_p)^2$$

(2.30)

Where V_{in} is the input voltage, and the loop gain of the circuit is given by $g_{m1}g_{m2}R_S R_p$. According to (2.30), the distortion of the circuit can cancel completely for specific values of the loop gain. This causes the large difficulty to maintain over a wide range of transistor variables.

Q-Enhanced Technique and Stability Analysis

Since the basic concept in the Q-enhanced LC filter is to use lossy LC tank, it is necessary to implement a loss compensation to boost the filter quality factor incorporating negativeconductance. Negative conductance g_{mF} realizes the required negative resistance to compensate for the loss in the tank. The effective quality factor [6] of the filter at the resonant frequency can be shown to be

$$Q_{en} = \frac{Q_0}{1 - g_{mF}R_p}$$

Where Q_0 is the base quality factor of the LC tank, which is dominated by the equivalent inductor. Theoretically it can be set as high as desired with appropriate g_{mF}. Indeed, the filter core can be tuned to oscillate if negative transconductance is sufficiently large, i.e., greater than $1/R_p$.

Additionally, the main problem is that the use of shunt feedback by M2 to compensate the loss resistance of the active inductor can result in potential instability depending on the filter terminating impedances yet. In order to make sure that the circuit is stable, the poles of the circuit must be in the left half-plane [16], [17]. In this condition, according to (1), using closed-loop analysis, the circuit will be stable provided that

$$g_{m1} < \frac{(C_{gs2} + C_{gd1})g_{o2}}{C_{gd2}} + g_{m2} + g_{oc}$$

(2.32)

Simultaneously it must be ensured that the magnitude of the input reflection coefficient is less than unity i.e. $|S_{11}| < 1$. Due to the stability problem, we should determine the reasonable transconductance g_{m1} and g_{m2} in order that the trade-offs between noise, Q enhancement and stability will satisfy the requirements of the communication systems.

Design of The RF Filter and Its Measured Results

Circuit Design

The complete prototype circuit of the proposed second-order RF bandpass filter based on the active inductor topology is shown in Figure 30. This circuit consists of three different stages, including two differential high Q-enhancement active inductors, negative impedance and buffers. Common-drain transistors M11, M13 and M12, M14 are employed for the output buffer stages. This common drain configuration can offer to minimize the loading effect and output impedance matching. M1, M3, M5 and M2, M4, M6 construct LC-resonant circuit which is made up of the active inductor respectively. Note that the transistor M5 and M6 are respectively used to amplify the signal of shunt feedback in the active inductor topology in order to boost the impedance of active inductors. M7, M8 and M9, M10 consisting of unbalanced cross-coupled pairs are employed not only to produce negative resistance for canceling the inductor loss, but also increase linearity of the filter when the signal is large. The transistors and capacitors are sized to optimize gain in the passband, noise figure, and linearity. Transistors M1, M2 have a length/width ratio of 2um/0.18um, M3, M4 have 4um/0.18um, M5, M6 have 20um/0.18um, and M7, M8 have 0.4um/0.2um and M9, M10 have 0.3um/0.18um. For the output buffers, transistors M11, M12 have a length/width ratio of 3um/0.18um, and M13, M14 have 2um/0.18um. The input capacitance is about 120ff. The DC bias current I_{Q1} and I_{Q2} can be used to tune the Q of the active inductors and the transconductance of the crosscoupled pairs. V_b and V_c are bias voltages which are used for DC operating state of the filter. The DC bias currents I_{bia1} and/or I_{bia2} can be adjusted to tune the center frequency of the circuit and also change the Q of the inductance in Fig. 3.

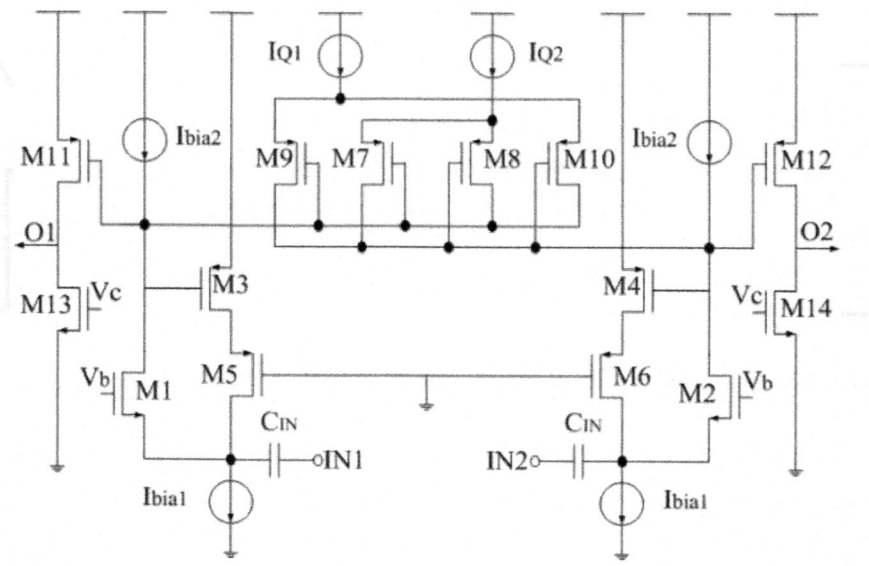

Figure 30: The fully Q-enhancement bandpass filter.

Measured Results

The circuit is fabricated in 0.18-um UMC-HJTC CMOS process through the educational service. The die photograph of the fabricated circuit is shown in Figure 31. To ensure the fully differential operation, a symmetrical layout is used for the design. The total chip area is $0.7 \times 0.75 \text{mm}^2$ including the pads, where the active area occupies only $0.15 \times 0.2 \text{mm}^2$.

The two-port S-parameter measurements were made with the vector network analyzer Agilent E8363B. Noise measurements were made with a spectrum analyzer equipped with power measurement software and a noise source. The 1-dB compression point measurements were made with a spectrum analyzer and a power meter. The measured RF bandpass filter forward transmission response, S21, is shown in Figure 32, Figure 33 and Figure 34, respectively. Figure 32 shows the passband center frequency is 1.92GHz and 3-dB bandwidth is about 28MHz. The maximum gain in the passband is about 11.64dB and the input return loss, S11 is -14.67dB in Figure 32. In Figure 33, the center frequency is about 2.44GHz and 3-dB bandwidth is about 60MHz. The maximum gain in the passband is about 5.99dB. Moreover, the S21 at about center frequency 3.82GHz is about 12dB and return loss S11 is about -29dB as shown in Figure 34.

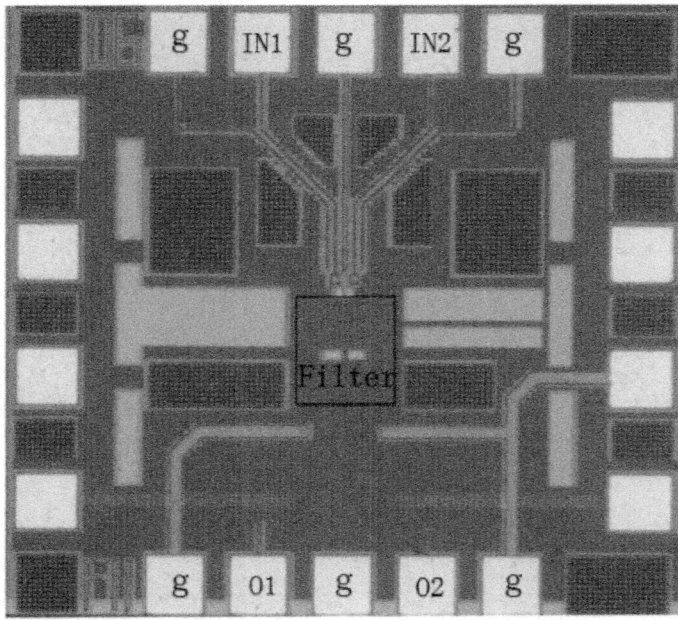

Figure 31: Photomicrograph of the Q-enhanced RF bandpass filter.

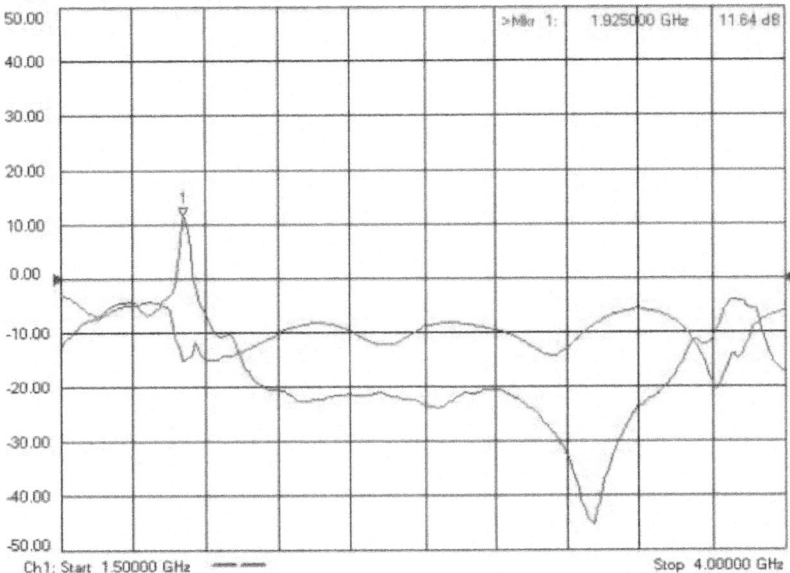

Figure 32: Measured bandpass filter insertion loss S21 and return loss S11 at center frequency about 1.92GHz.

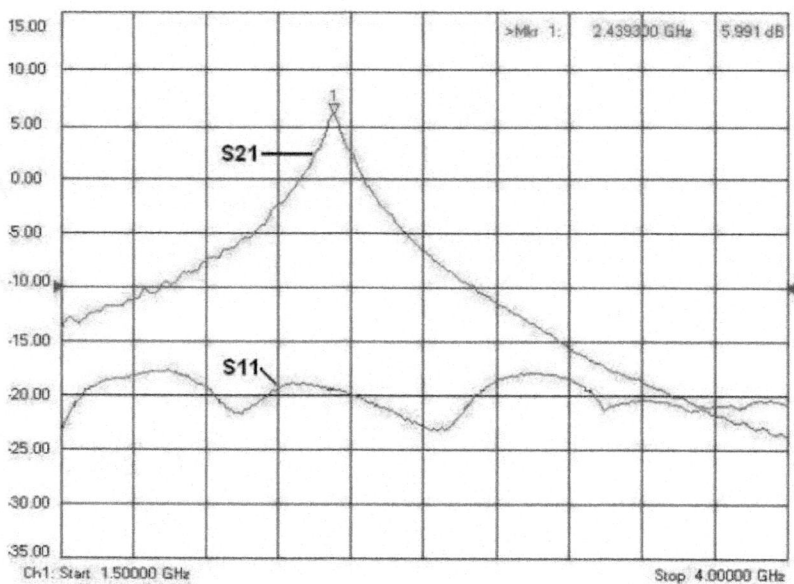

Figure 33: Measured bandpass filter insertion loss S21 and return loss S11 at center frequency about 2.44GHz.

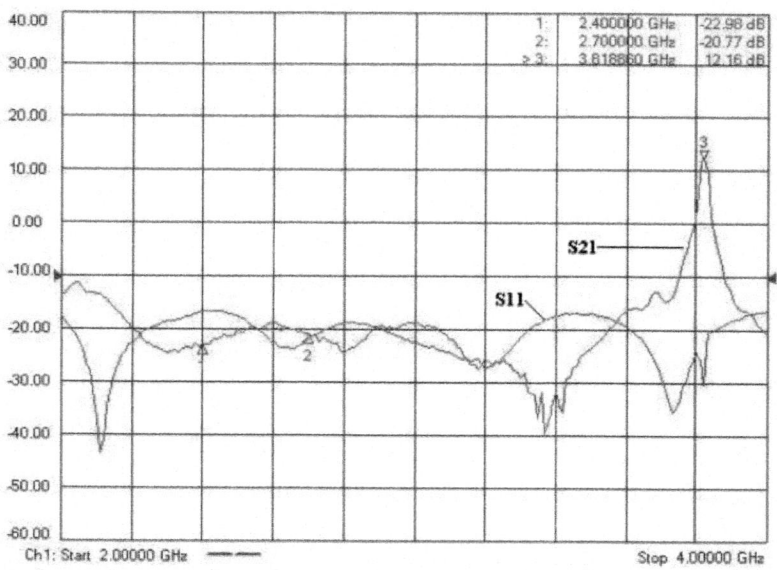

Figure 34: Measured bandpass filter insertion loss S21 and return loss S11 at the center frequency about 3.82GHz

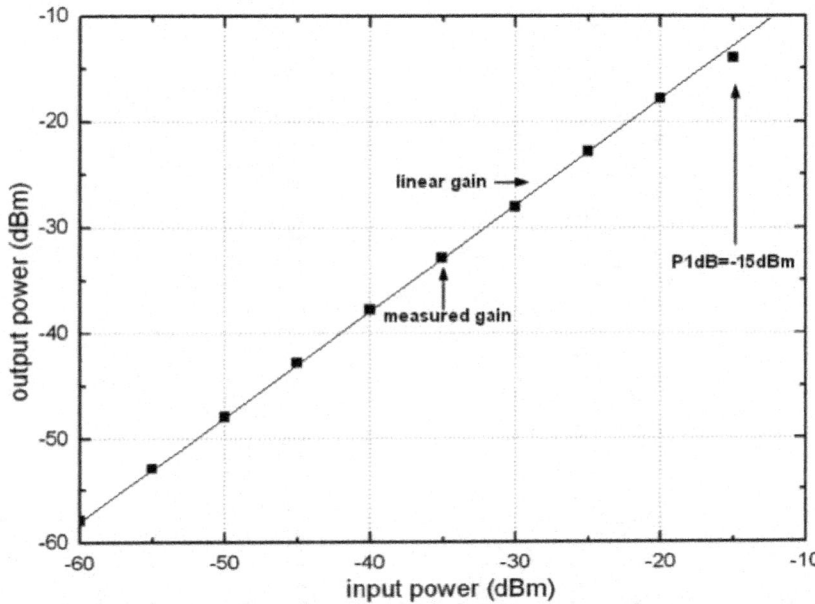

Figure 35: P1dB measurement at center frequency about 2.44GHz.

A measurement to the input 1dB compression point of the circuit can be obtained by sweeping the input power to the tank and measuring the output power. As the input power is increased, the input impedance presented by the Q-enhanced active inductor tanks begins to drop due to nonlinear effects, which can be observed when the output power no longer depends on the input power in a linear fashion as shown in Figure 35. The measured bandpass filter P1dB input power compression point is -15dBm at the center frequency about 2.44GHz passband. The noise figure of 18dB was also measured by disconnecting the input signal. The RF filter has wide-tuning range from the center frequency about 1.92GHz to 3.82GHz when the DC voltage sources of the controlled bias currents Ibia1 and/ or Ibia2 are adjusted from 0.5 to 1.5V or vice versa. The noise figure evaluated in each band gives the following results: 15dB for center frequency 1.92GHz, 18dB for center frequency about 2.44GHz, 20dB for center frequency about 3.82GHz. Furthermore, 1-dB compress point is about -17dBm, -15dBm, -18dBm respectively.

Table 2: Comparison of the RF bandpass filters Performance

Ref.	[8]	[9]	[10]	[11]	This work
Process	0.25um-CMOS	0.5-Si-SOI	0.18um-CMOS	0.18um-CMOS	0.18um-CMOS
Die area	3.5mm²	2.5mm²	2.25mm²	0.81mm²	0.53mm²
N-orders	6	2	3	4	2
f_{center}	2.14GHz	2.5GHz	2.36GHz	2.03GHz	2.44GHz
-3dB Bandwidth	60MHz	70MHz	60MHz	130MHz	60MHz
Tuning range	-	250MHz	-	60.9MHz	1900MHz
Noise figure	19dB	6dB	18 dB	15 dB	18 dB
Mid-band gain	0dB	14dB	-1.8dB	0dB	6dB
Supply voltage	2.5V	3V	1.8V	1.8V	1.8V
FOM	72	82	78	77	81

The summary of the measured performance and the comparisons of the performance among the fabricated RF filters in CMOS and other process is given in Table 2. A figure of merit [18] (FOM) which allows comparison between other RF filters in silicon is given as

$$FOM = \frac{N \cdot P_{1dBW} \cdot f_{center} \cdot Q_{filter}}{P_{DC} \cdot NF}$$

(2.33)

where N is the number of poles, P_{1dBW} is the inband 1-dB compression point in Watt, f_{center} is the center frequency, Q_{filter} is the ratio of the center frequency and the 3-dB bandwidth, P_{DC} is the DC power dissipation in Watt, and NF is the noise figure (not in dB). It has shown from Table II that the filter presented in this work achieves a good FOM with higher quality factor and gain in the passband, and the tuning range is the largest, and the chip area is the smallest.

CONCLUSION

The design and implementation of tunable RF bandpass filter in 0.18um CMOS process have been introduced and verified, which demonstrate that the RF bandpass filter can achieve high quality factor and large tuning range from 1.92GHz to 3.82GHz. Although the noise and linearity of the proposed active inductors are inferior to passive ones, the smallest chip area, and the largest tenability make them apply to the multi-band on-chip wireless systems in future.

REFERENCES

1. J. Ryynäen, S. Lindfors, et al. Integrated Circuits for Multi-Band Multi-Mode Receivers. IEEE Circuits and System Magazine. 2006, 6(2): 5-16.

2. Ahmed N. M.,Edgar S. S., Jose S. M., A fully balancedpseudo-differential OTA with common-Mode Feedforward and Inherent common-mode feedback detector IEEE J.Of Solid-State Circuits, vol.38,No.4, p.663-668 2003.

3. L. E. Aguado, K. K. Wong, et al. Coexistence Issues for 2.4 GHz OFDM WLANs. 3G Mobile Communication Technologies, London, 2002: 400–404.

4. D. Chamla, A. Kaiser. A Gm-C Low-pass Filter for Zero-IF Mobile Applications with a Very Wide Tuning Range. IEEE Journal of Solid-State Circuits. 2005, 40(7): 1443-1450.

5. A. D. Sanchez, R. Anglulo, J., et al. A CMOS Four Quadrant Current/ Transconductance Multiplier, Analog Integrated Circuits Signal Process. 1999, 19(2): 163-168.

6. R. G. Carvajal, et al. The Flipped Voltage Follower: A Useful Cell for Low-Voltage LowPower Circuit Design, IEEE Transactions on Circuits and System-I: Regular Paper. 2005, 52(7): 1276-1289.

7. B. Georgescu, H. Pekau, J. Haslett and J. Mcrory, Tunable coupled inductor Qenhancement for parallel resonant LC tanks, IEEE Trans. Circuit and System-II: analog and digital signal processing, vol. 50, pp705-713, Oct. 2003.

8. T.H. Lee, The design of CMOS radio-frequency integrated circuits, U.K.: Cambridge Univ. Press, pp.390-399, 2004.

9. W.B.Kuhn, F. W. Stephenson, and A. Elshabini-Riad, A 200MHz CMOS Q-enhanced LC bandpass filter, IEEE J. Solid-State Circuits, vol. 31, pp.1112-1122, Oct. 1996.

10. W. B. Kuhn, D. Nobbe, D. Kelly, A. W. Orsborn, Dynamic range performance of on-chip RF bandpass filters, IEEE Trans. Circuit and Systems-II:analog and digital signal processing, vol. 50, pp. 685-694, Oct. 2003.

11. S. Bantas, Y. Koutsoyannopoulos, CMOS active-LC bandpass filters with coupled inductor Q-enhancement and center frequency tuning, IEEE Trans. Circuits and Systems-II: express briefs, vol. 51, pp.69-77 Feb. 2004.

12. S.Pipilos, Y.P. Tsividis, J. Fenk, and Y. Papanaos, A Si 1.8 GHz RLC filter with tunable center frequency and quality factor, IEEE J. Solid-State Circuits, vol. 31, pp. 1517- 1525, Oct. 1996.

13. F. Dulger E. S. Sinencio and J. Silva-Martinez, A 1.3V 5mW fully integrated tunable bandpass filter at 2.1GHz in 0.35um CMOS, IEEE J. Solid-State Circuits, vol. 38 pp. 918-927, June 2003.

14. W.B. Kuhn, A. Elshabini-Riad, and F. W. Stephenson, Center-tapped spiral inductors for monolithic bandpass filters, Electron. Lett., vol. 31, pp.625-626, Apr. 1995.

15. F. Krummenacher, G. V. Ruymbeke, Integrated selectivity for narrow-band FM IF systems, IEEE J. Solid-State Circuits, vol. SC-25, pp.757-760, June 1990.

16. T. Soorapanth, S. S. Wong, A 0dB IL 2140 ± 30 MHz bandpass filter utilizing Qenhanced spiral inductors in standard CMOS, IEEE J. Solid-State Circuits vol.37, pp. 579-586, may, 2002.

17. A. Tasic, W.A. Serdijn, J.R. Long. Adaptive multi-standard circuits and systems for wireless communications. IEEE Magazine On Circuits and Systems, vol. 6, no. 1, pp. 29-37, Quarter, 2006.

18. Y. Satoh, O. Ikata, T. Miyashita, and H. Ohmori, RF SAW filters Chiba Univ., Japan, 2001

19. Online.. Available: http://www.usl.chiba-u.ac.jp/ken /Symp2001/ PAPER/SATOH.PDF.

20. A. Tasic, S. Lim, W.A. Serdijn, J.R. Long. Design of adaptive multimode RF front-end circuits. IEEE J. of Solid-State Circuits, vol. 42, pp. 313-322, Feb. 2007.

21. MBC13916 General purpose SiGe: C RF Cascade Amplifier. Motorola, Data Sheet.

22. S. Li, N. Stanic, Y. Tsividis. A VCF loss-control tuning loop for Q-enhanced LC filters. IEEE Trans. On Circuits and Systems-II: express briefs, vol. 53, pp. 906-910, Sep. 2006.

23. W. B. Kuhn, D. Nobe, N. Kely, A. W. Orsborn. Dynamic range performance of on-chip RF bandpass filters. IEEE Trans. On Microwave Theory and Techniques, vol. 50, pp. 685- 694, Oct. 2003.

24. B. Bantas, Y. Koutsoyannopoulos. CMOS active-LC bandpass filters with coupledinductor Q-enhancement and center frequency tuning. IEEE Trans. Circuits and Systems-II: express briefs, vol. 51, pp. 69-76, Feb. 2004.

25. T. Soorapanth S. S. Wong. A 0-dB IL, 2140±30MHz bandpass filter utilizing Q-enhanced spiral inductors in standard CMOS. IEEE J. of Solid-State Circuits, vol. 37, 579-586, May 2002.

26. X. He, W. B. Kuhn. A 2.5GHz low-power, high dynamic range, self-tuned Q-enhanced LC filter in SOI. IEEE J. of Solid-State Circuits, vol. 40, pp.1618-1628, Aug. 2005.

27. J. Kulyk, J. Haslett. A monolithic CMOS 2368±30MHz transformer based Q-enhanced series-C coupled resonator bandpass filter. IEEE J. of Solid-State Circuits, vol. 41, 362-374, Feb. 2006.

28. B. Georgescu, I. G. Finvers, F. Ghannouchi. 2 GHz Q-enhanced active filter with low passband distortion and high dynamic range. IEEE J. of Solid-State Circuits, vol. 41, pp.2029-2039, Sep. 2006.

29. Y.Cao, R. A. Groves, X. Huang et. al. Frequency-independent equivalent-circuit model for on-chip spiral inductors. IEEE J. of Solid-State Circuits, vol. 38, 419-426, Mar. 2003.

30. W. Sansen. Distortion in elementary transistor circuits. IEEE Trans. On Circuits and Systems-II: express briefs, vol. 46, 315-325, Mar. 1999.

31. Z. Gao, M. Yu, Y. Ye, and J. Ma. A CMOS RF tuning wide-band bandpass filter for wireless applications. In Proc. IEEE Int'l Conf. SOC, pages 79–80, Spet. 2005.

32. G. Groenewold. Noise and group delay in active filters. IEEE Trans. On Circuits and Systems-I: regular papers, vol. 54, pp. 1471-1480, July, 2007.

33. P. R. Gray and R. G. Meyer, Analysis and Design of Analog Integrated Circuits, 3rd ed. New York: Wiley, 1993.

34. Robert W. Jackson. Rollett Proviso in the stability of linear Microwave circuits-A tutorial. IEEE Trans. on Microwave Theory and Techniques, vol. 54, pp. 993-1000, Mar. 2006.

35. K.T. Christensen, T. H. Lee, E. Bruun. A high dynamic range programmable CMOS Front-end filter with tuning range from 1850-2400 MHz. Norwell, MA: Kluwer, 2005.

36. A.Karsilayan and R. Schaumann A High-Frequency High-Q CMOS Active Inductor with DC Bias Control, Proc. IEEE Midwest Symp. Circ. Syst. (MWSCAS), Aug. 2000.

37. Y. Wu, M. Ismail, and H. Olsson, A novel fully differential inductorless RF bandpass filter in Proc. IEEE Int. Symp. Circuit and System (ISCAS), Geneva, Switzerland, May 2000, pp. 149-152.

Chapter 8

TRENDS OF THE OPTICAL WIRELESS COMMUNICATIONS

Juan-de-Dios Sánchez- López[1], Arturo Arvizu M[2], Francisco J. Mendieta[2] and Iván Nieto Hipólito[1]

[1]Autonomus University of Baja California,

[2]Cicese Research Center México

INTRODUCTION

The Optical Wireless Communications (OWC) is a type of communications system that uses the atmosphere as a communications channel. The OWC systems are attractive to provide broadband services due to their inherent wide bandwidth, easy deployment and no license requirement. The idea to employ the atmosphere as transmission media arises from the invention of the laser. However, the early experiments on this field did not have any baggage of technological development (like the present systems) derived from the fiber optical communications systems, because like this, the interest on them decreased (Willebrand, 2002). At the beginning of the last century, the OWC systems have attracted some interest due to the advantages mentioned above. However, the interaction of the electromagnetic waves with the atmosphere at optical frequencies is stronger than that corresponding at microwave. (Wheelon, 2002).

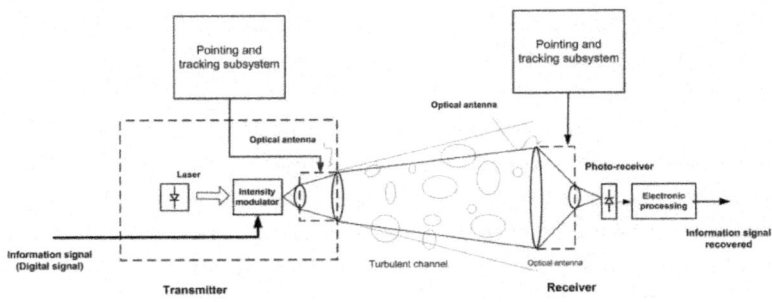

Figure 1: Model of a typical atmospheric optical communications link.

The intensity of a laser beam propagating through the atmosphere is reduced due to phenomena such as scattering and molecular absorption, among other (Willebrand, 2002). The changes in the refractive index of the atmosphere due to optical turbulence affect the quality of laser beam through distortion of its phase front and random modulation of its optical power (Zsu, 2002). Also the presence of fog may completely prevent the passage of the optical beam that leads to a no operational communications link (Kedar, 2003).

The figure 1 shows the block diagram of an OWC communications system (also called Free Space optic communications system or FSO) (Zsu, 2002). The information signal (analog or digital) is applied to the optical transmitter to be sent through the atmosphere using an optical antenna. At the receiver end the optical beam is concentrated, using an optical antenna, to the photo-detector sensitive area, which output is electrically processed in order to receiver the information signal.

IMPORTANT ACCESS TECHNOLOGIES (FIRST AND LAST MILE)

In the past decades, the bandwidth of a single link in the backbone of the networks has been increased by almost 1000 times, thanks to the use of wavelength division multiplexing (WDM) [Franz, 2000]. The existing fiber optic systems can provide capabilities of several gigabits per second to the end user. However, only 10% of the businesses or offices, have direct access to fiber optics, so most users who connect to it by other transmission technologies which use copper cables or radio signals, which reduces the throughput of these users. This is a bottleneck to the last mile (Zsu, 2002).

While there are communication systems based on broadband DSL technology or cable modems, the bandwidth of these technologies is limited when compared against the optical fiber-based systems (Willebrand, 2002). In the other hand, the RF systems using carrier frequencies below the millimeter waves can not deliver data at rates specified by IEEE 802.3z Gbit Ethernet. Rates of the 1 Gbps and higher can only be delivered by laser or millimeter-wave beams. However, the millimeter wave technology is much less mature than the technology of lasers (Willebrand, 2002), which leaves the optical communications systems as the best candidates for this niche market. Therefore, the access to broadband networks based on optical communications may be accomplished through passive optical networks (or PON's, which are based on the use of fiber optics) or via optical wireless communication systems (Qingchong, 2005).

The optical wireless communications industry has experienced a healthy growth in the past decade despite the ups and downs of the global economy.

This is due to the three main advantages over other competing technologies. First, the wireless optical communications cost is on average about 10% of the cost of an optical fiber system (Willebrand, 2002). It also requires only a few hours or weeks to install, similar time to establish a radio link (RF), while installing the fiber optics can take several months. Second, OWC systems have a greater range than systems based on millimeter waves. OWC systems can cover distances greater than a kilometer, in contrast with millimeter-wave systems that require repeaters for the same distance. In addition, millimeter wave systems are affected by rain, but the OWC systems are affected y fog, which makes complementary these transmission technologies (Qingchong, 2005). Finally, this type of technology as opposed to radio links, does not require licensing in addition to not cause interference.

Applications of the OWC systems

Optical wireless communications systems have different applications areas:

a.   Satellite networks The optical wireless communications systems may be used for in satellite communication networks, satellite-to-satellite, satellite-to-earth (Hemmati et al, 2004).

b.   Aircraft In applications satellite to aircraft or the opposite (Lambert et al, 1995).

c.   Deep Space In the deep space may be used for communications between spacecraft – to – earth or spacecraft to satellite. (Hemmati et al, 2004).

d.   Terrestrial (or atmospheric) communications In terrestrial links are used to support fiber optic, optical wireless networks "wireles optical networks (WON)" last mile link, emergency situations temporary links among others (Zsuand & Kahn, 2002).

Each application has different requirements but this book chapter deals primarily with terrestrial systems.

Basic Scheme of OWC Systems Communications

Optical communications receivers can be classified into two basic types. (Gagliardi & Karp, 1995): non-coherent receivers and coherent receivers. Noncoherent detect the intensity of the signal (and therefore its power). This kind of receivers is the most basic and are used when the information transmitted is sent by the variations in received field strength. On the other hand are coherent receivers, in which the received optical field is mixed with the field generated by a local optical oscillator (laser) through a beam combiner or coupler, and the resulting signal is photo-detected.

Noncoherent Optical Communications Systems

The commercially deployed OWC systems use the intensity modulation (IM) that is converted into an electrical current in the receiver by a photodetector (usually are a PIN diode or an avalanche photo diode (APD)) which is known as direct detection (DD). This modulation scheme is widely used in optical fiber communications systems due to its simplicity.

In IM-DD systems, the electric field of light received, E_s is directly converted into electricity through a photoreceiver, as explained above. The photocurrent is proportional to the square of E_s and therefore the received optical power P_r, i.e.:

$$i(t) = \frac{e\eta}{h\nu} E_s^2(t) \tag{1}$$

where e is the electronic charge, η is the quantum efficiency, h is Planck's constant, ν is the optical frequency. The block diagram of the system is shown in Figure 2.

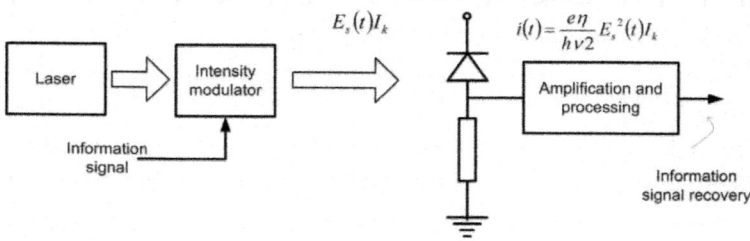

Figure 2: Block diagram using an optical communication system of intensity modulation and direct detection (noncoherent).

The optical direct detection can be considered as a simple process of gathering energy that only requires a photodetector placed in the focal plane of a lens followed by electronic circuits for conditioning the electrical signal derived from the received optical field (Franz & Jain, 2000).

Coherent Optical Communications Systems

In analog communications in the radio domain [Proakis, 2000, Sklar, 1996], the coherent term is used for systems that recover the carrier phase. In coherent optical communications systems, the term "coherent" is defined in a different way: an optical communication system is called coherent when doing the mixing of optical signals (received signal and the signal generated locally) without necessarily phase optical carrier recovered [Kazovsky, 1996]. Even if it does not use the demodulator carrier recovery but envelope detection, the system is called coherent optical communication system due to the mixing operation of the optical signals. In turn, the coherent receivers can be classified into two types: asynchronous and synchronous. They are called synchronous when the tracking and recovering of the carrier phase is performed and asynchronous when is not performed the above mentioned process. The asynchronous receivers typically use envelope detection (Kazovsky, 1996), (Franz & Jain, 2000) Figure 3 shows the basic structure of a communications system with digital phase modulation and coherent detection. The output current of the photodetectors array is:

$$i(t) = \Re \frac{E_s^2(t)}{2} + \Re \frac{E_{LO}^2(t)}{2}$$
$$+ \Re \sqrt{E_s(t)E_{LO}} \cos\left\{[\omega_{LO} - \omega_s]t + \phi_{LO} - \phi_s\right\} \tag{2}$$

where $\Re = en/h\nu$ is the responsivity, E_{LO} is the electric field generated by the laser that operates as a local oscillator, ω_{LO} is the frequency of the local oscillator and ω_s is the carrier frequency of the optical received signal φ_{LO} is the phase of the carrier signal received, and φ_s is the carrier phase of the received optical signal. The coherent mixing process requires that the local beam to be aligned with the beam received in order to get efficient mixing. This can be implemented in two different ways; if the frequency of signal and local oscillator are different and uncorrelated the process is referred to as heterodyne detection (Fig. 4) (Osche, 2002); if the frequencies of the signal and local oscillator are the same and are correlated, is called homodyne detection (Fig. 5) (Osche, 2002).Due to the process of mixing, coherent receivers are theoretically more sensitive than direct detection receivers (Kazovsky, 1996).

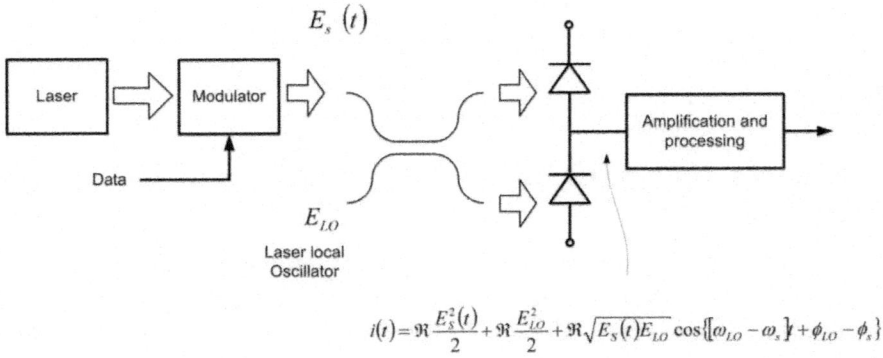

Figure 3: Optical Communication System with coherent detection.

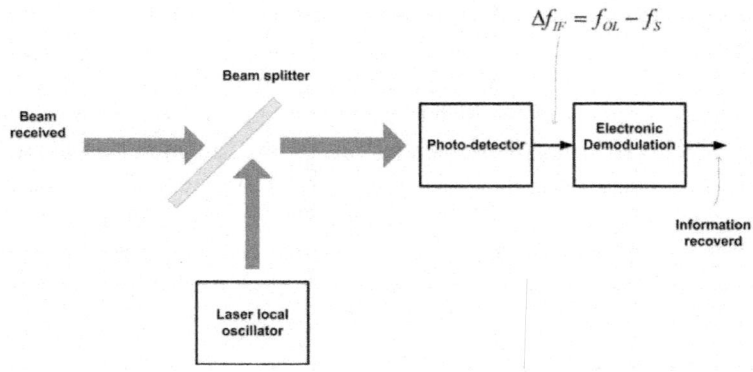

$$\Delta f_{IF} = f_{OL} - f_S$$

Figure 4: Optical heterodyne receiver.

In terms of sensitivity, the coherent communications systems with phase modulation, theoretically have the best performance of all (e.g. BPSK is about 20 dB better than OOK). Sensitivity is the number of photons per bit required to get a given probability of error (Kazovsky 1996).

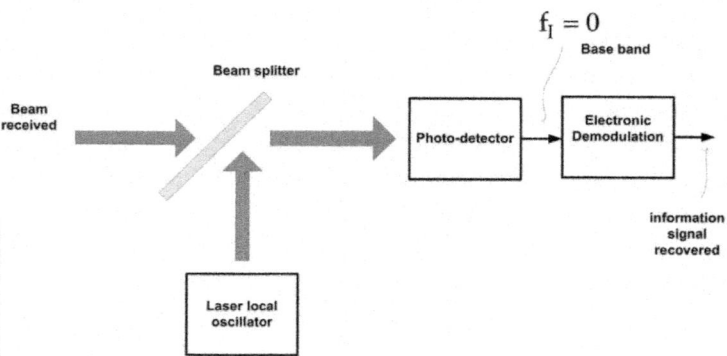

$$f_I = 0$$

Figure 5: Optical homodyne receiver.

Advantages of Optical Communications Systems with Coherent Detection

As mentioned previously the coherent optical communications systems have better performance than incoherent optical communications systems and may be used the phase, amplitude and frequency and state of polarization (SOP) of the optical signal allowing various digital modulation formats of both amplitude, phase and SOP combination. However, the coherent detection systems are expensive and complex (Kazovsky, 1996), (Ryu, 1995) and require control mechanisms or subsystems of the state of polarization of the received signal with the optical signal generated by local oscillator (laser). Moreover, homodyne optical communications systems require coherent phase recovery of the optical carrier, and usually this is done through optical Phase Lock Loop (OPLL), Costas loop or other sinchronization technique, which increases the complexity of these systems.

OPTICAL AND OPTOELECTRÓNIC COMPONENTS

Devices such as the laser diodes, high-speed photo-receivers, optical amplifiers, optical modulators among others are derived of about thirty years of investigation and development of the fiber optics telecommunications systems. These technological advances has made possible the present OWC systems. Additionally, OWC systems have been benefited by the advances in the telescopes generated by the astronomy.

Optical Sources for Transmitters

In modern optical wireless communications, there are a variety of light sources for use in the transmitter. One of the most used is the semiconductor laser which is also widely used in fiber optic systems. For indoor environment applications, where the safety is imperative, the Light Emitter Diode (LED) is prefered due to its limited optical power. Light emitting diodes are semiconductor structures that emit light. Because of its relatively low power emission, the LED's are typically used in applications over short distances and for low bit rate (up to 155Mbps). Depending on the material that they are constructed, the LED's can operate in different wavelength intervals. When compared to the narrow spectral width of a laser source, LEDs have a much larger spectral width (Full Width at Half Maximun or FWHM). In Table 1 are shown the semiconductor materials and its emission wavelength used in the LED's (Franz et al, 2000).

Table 1: Material, wavelength and energy band gap for typical LED

Material	Wavelength Range (nm)
AlGaAs	800 – 900
InGaAs	1000 – 1300
InGaAsP	900 – 1700

Laser

The laser is an oscillator to optical frequencies which is composed by an optical resonant cavity and a gain mechanism to compensate the optical losses.

Semiconductor lasers are of interest for the OWC industry, because of their relatively small size, high power and cost efficiency. Many of these lasers are used in optical fiber systems, there is no problem of availability. Table 2 summarize the materials commonly used in semiconductor lasers (Agrawal, 2005).

Table 2: Materials used in semiconductor laser with wavelengths that are relevant for FSO

Material	Wavelength Range (nm)
GaAlAs	620 - 895
GaAs	904
InGaAsP	1100 – 1650
	1550

Photodetectors

At the receiver, the optical signals must be converted to the electrical domain for further processing, this conversion is made by the photo detectors. There are two main types of photodetectors, PIN diode (Positive-Intrinsic-Negative) and avalanche photodiode" avalanche photodiode (APD) (Franz et al, 2000). The main parameters that characterize the photodetectors in communications are: spectral response, photosensitivity, quantum efficiency, dark current, noise equivalent power, response time and bandwidth (Franz et al, 2000). The photodetection is achieved by the response of a photosensitive material to the incident light to produce free electrons. These electrons can be directed to form an electric current when applied an external potential.

Pin Photodiode

This type of photodiodes have an advantage in response time and operate with reverse bias. This type of diode has an intrinsic region between the PN materials, this union is known as homojunction. PIN diodes are widely used in telecommunications because of their fast response. Its responsivity, i.e. the ability to convert optical power to electrical current is function of the material and is different for each wavelength. This is defined as:

$$\Re = \frac{\eta e}{h \nu} \quad [A/W]$$

(3)

Where η is the quantum efficiency, e is the electron charge (1.6×10^{-19} C), h is Planck's constant (6.62×10^{-34} J) and ν is the frequency corresponding to the photon wavelength.

InGaAs PIN diodes show good response to wavelengths corresponding to the low attenuation window of optical fiber close to 1500nm. The atmosphere also has low attenuation into regions close to this wavelength.

Avalanche Photodiode

This type of device is ideal for detecting extremely low light level. This effect is reflected in the gain M:

$$M = \frac{I_G}{I_p}$$

(4)

I_G is the value of the amplified output current due to avalanche effect and I_p is the current without amplification. The avalanche photo diode has a higher output current than PIN diode for a given value of optical input power, but the noise also increases by the same factor and additionally has a slower response than the PIN diode (see table 3).

Table 3: Characteristics of photo detectors used in OWC systems

Material and Structure	Wavelength (nm)	Responsivity (A/W)	Gain	Rise time
PIN. Silicon	300 – 1100	0.5	1	0.1-5 ns
PIN InGaAs	1000 – 1700	0.9	1	0.01-5 ns
APD Germanium	800 – 1300	0.6	10	0.3-1 ns
APD InGaAs	1000 – 1700	0.75	10	0.3 ns

Table 3 shows some of the materials and their physical properties used to manufacture of photo-detectors (Franz et al, 2000).

Optical Amplifiers

Basically there are two types of optical amplifiers that can be used in wireless optical communication systems: semiconductor optical amplifier (SOA) and amplifier Erbium doped fiber (EDFA). Semiconductor optical amplifiers (SOA) have a structure similar to a semiconductor laser, but without the resonant cavity. The SOA can be designed for specific frequencies. Erbium-doped fiber amplifiers are widely used in fiber optics communications systems operating at wavelenghts close to 1550 nm. Because they are built with optical fiber, provides easy connection to other sections of optical fiber, they are not sensitive to the polarization of the optical signal, and they are relatively stable under environment changes with a requirement of higher saturation power that the SOA.

Optical Antennas

The optical antenna or telescope is one of the main components of optical wireless communication systems. In some systems may have a telescope to the transmitter and one for the receiver, but can be used one to perform both functions. The transmitted laser beam characteristics depend on the parameters and quality of the optics of the telescope. The various types of existing telescopes can be used for optical communications applications in free space. The optical gain of the antennas depends on the wavelength used and its diameter (see equations 5, 40 and 41). The Incoherent optical wireless communication systems typically expands the beam so that any change in alignment between the transmitter and receiver do not cause the beam passes out of the receiver aperture. The beam footprint on the receiver can be determined approximately by:

$$D_f \approx \theta L \tag{5}$$

D_f is the footprint diameter on the receiver plane in meters, θ is the divergence angle in radians and L is the separation distance between transmitter and receiver (meters). The above approximation is valid considering that the angle of divergence is the order of milliradians and the distances of the links are typically over 500 meters.

FACTORS AFFECTING THE TERRESTRIAL OPTICAL WIRELESS COMMUNICATIONS SYSTEMS

Several problems arise in optical wireless communications because of the wavelengths used in this type of system (Osche, 2002). The main processes affecting the propagation in the atmosphere of the optical signals are absorption, dispersion and refractive index variations (Collet, 1970), (Goodman, 1985) (Andrews, 2005), (Wheelon, 2003). The latter is known as atmospheric turbulence. The absorption due to water vapor in addition with scattering caused by small particles or droplets or water (fog) reduce the optical power of the information signal impinging on the receiver (Willebrand, 2002). Because of the above mentioned previously, this type of communications system is suscpetible to the weather conditions prevailing in its operating enviroment. Figure 6 shows the disturbances affecting the optical signal propagation through the atmosphere.

Fog

Fog is the weather phenomenon that has the more destructive effect over OWC systems due to the size of the drops similar to the optical wavelengths used for communications links (Hemmati et al, 2004.). Dispersion is the dominant loss mechanism for the fog (Hemmati et al, 2004.). Taking into account to the effect over the visibility parameter the fog is classified as low (1-5 km), moderate (0.2-1 km) and dense (0.034 – 0.2 km). The attenuation due to visibility can be calculated using the following equation (Kim et al, 2000):

$$P_v = \exp\left[\frac{-3.9}{V}\left(\frac{\lambda}{0.55}\right)^m L \right]$$

(6)

Where V is the visibility [km], L is the propagation range and m is the size distribution for the water drops that form the fog.

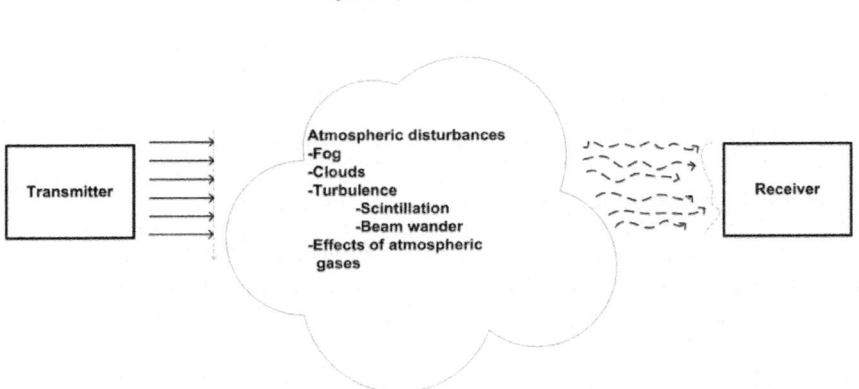

Figure 6: Optical link over a terrestrial atmospheric channel.

Rain

Other weather phenomena affecting the propagation of an optical signal is the rain, however its impact is in general negligible compared with the fog due to the radius of the drops (200µm - 2000µm) which is significantly larger than the wavelength of the light source OWC systems [Willebrand 2002].

Effects due to Atmospheric Gases. Dispersion and Absorption

The dispersion is the re-routing or redistribution of light which significantly reduces the intensity arriving into the receiver (Willebrand, 2002). The absorption coefficient is a function of the absorption of each of the particles, and the particle density. There absorbent which can be divided into two general classes: molecular absorbent (gas) ; absorbing aerosol (dust, smoke, water droplets).

Atmospheric Windows

The FSO atmospheric windows commonly used are found in the infrared range. The windows are in $0.72\mu m$ and $1.5\mu m$, and other regions of the absorption spectrum. The region of $0.7\mu m$ to $2.0\mu m$ is dominated by the absorption of water vapor and the region of $2.0\mu m$ to $4.0\mu m$ is dominated by the combination of water and carbon dioxide.

Aberrations Losses

These losses are due to the aberrations of the optical elements and can be expressed as:

$$L_{ab} = e^{-(k\sigma_a)^2}$$

(7)

$k=2\pi/\lambda$

σ_a=rms aberrations error

Atmospheric Attenuation

Describes the attenuation of the light traveling through the atmosphere due to absorption and dispersion. In general the transmission in the atmosphere is a function of link distance L, and is expressed in Beer's law as [Lambert et al, 1995]

$$L_{atm} = 10\log \tau \left[\frac{dB}{Km} \right]$$

(8)

And

$$\frac{I_d}{I_{Tx}} = \tau = \exp(-\gamma L)$$

(9)

I_d/I_{Tx} is the relationship between the intensity detected and the transmitted output intensity and γ is the attenuation coefficient. The attenuation coefficient is the addition of four parameters; the dispersion coefficients of molecules and aerosols, α and absorption coefficient, β of molecules and aerosols, each depending on the wavelength and is given by (Lambert et al 1995).

$$\gamma = \alpha_{molecule} + \alpha_{aerosol} + \beta_{molecule} + \beta_{aerosol}$$

(10)

Atmospheric Turbulence

Inhomogeneities in temperature and pressure variations of the atmosphere cause variations in the refractive index, which distort the optical signals that travel through the atmosphere. This effect is known as atmospheric turbulence.The performance of atmospheric optical communications systems will be affected because the atmosphere is a dynamic and imperfect media. Atmospheric turbulence effects include fluctuations in the amplitude and phase of the optical signal (Tatarski, 1970), (Wheelon, 2003). The turbulence-induced fading in optical wireless communication links is similar to fading due to multipaths experienced by radiofrequency communication links (Zsu, 2002). The refractive index variations can cause fluctuations in the intensity and phase of the received signal increasing the link error probability.

As mentioned briefly above, the heating of air masses near the earth's surface, which are mixed due to convection and wind generates atmospheric turbulence. These air masses have different temperatures and pressure values which in turn leads to different refractive index values, affecting the light traveling through them. The atmospheric turbulence has important effects on a light beam especially when the link distance is greater than 1 km (Zsu, 1986). Variations in temperature and pressure in turn cause variations in the refractive index along the link path (Tatarski, 1971) and such variations can cause fluctuations in the amplitude and phase of the received signal (known as flicker or scintillation) (Gagliardi, 1988). Kolmogorov describe the turbulence by eddies, where the larger eddies are split into smaller eddies without loss of energy, dissipated due to viscosity (Wheelon, 2003, Andrews, 2005), as shown in Figure 7. The size of the eddies ranges from a few meters to a few millimeters, denoted as outer scale L_0, and inner scale, l_0, respectively as shown in Figure 7 and eddies or inhomogeneities with dimensions that are between these two limits are the range or inertial subrange (Tatarski, 1971).

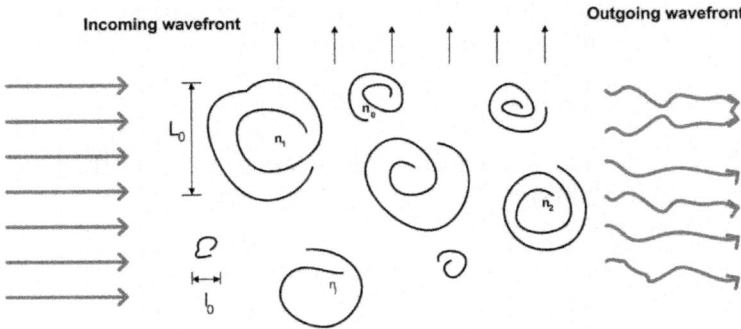

Figure 7: Turbulence model based on eddies according to the Kolmogorov theory.

A measure of the strength of turbulence is the constant of the structure function of the refractive index of air, C_n^2, which is related to temperature and atmospheric pressure by (Andrews, 2005):

$$C_n^2 = \left(79 \times 10^{-6}\frac{P}{T}\right)^2 C_T^2$$

$$(11)$$

Where P is the atmospheric pressure in millibars, T is the temperature in Kelvin degrees and $C_T{}^2$ is the constant of the structure function. In short intervals, at a fixed propagation distance and a constant height above the ground can be assumed that $C_n{}^2$ is almost constant, (Goodman, 1985). Values of $C_n{}^2$ of 10-17 m-2/3 or less are considered weak turbulence and values up to 10-13m-2/3 or more as strong turbulence (Goodman, 1985). We can also consider that in short time intervals, for paths at a fixed height, $C_n{}^2$ is constant (the above for horizontal paths). $C_n{}^2$ varies with height (Goodman, 1985).

Another measure of the turbulence is the Rytov variance, which relates the structure constant of refractive index with the beam path through the following equation:

$$\sigma_R^2 = 1.23C_n^2 k^{7/6}L^{11/6}$$

$$(12)$$

where λ is the wavelength, L is the distance from the beam path and $k=2\pi/\lambda$. An optical light beam is affected by turbulence in different ways: variations in both intensity and amplitude, phase changes (phase front), polarization fluctuations and changes on the angle of arrival.

Intensity and Amplitude Fluctuations

The atmospheric turbulence affects the amplitude and phase of the optical signal that propagates through the medium in two points separated by a distance r, and can be described by the following equation according to the Rytov method for solving Maxwell's equations (Goodman, 1985):

$$U(\bar{r}) = U_0(\bar{r})\exp(\psi(\bar{r}))$$

$$(13)$$

where $U_0(r)$ is the undisturbed field. The complex phase perturbation can be written (Andrews, 2005):

$$\psi_1(\bar{r}) = \chi + iS_1$$

$$(14)$$

Or

$$\psi_1(\bar{r}) = \ln\left(\frac{A}{A_0}\right) + i(S - S_0)$$

$$(15)$$

where χ is the logarithm of the amplitude A and S is the phase of the field U(r) and A0 and S0 are the amplitude and phase without disturbing respectively. This analysis is done based on the Rytov approximation and shows that the irradiance (or intensity) fluctuations follow a lognormal distribution due to that the logarithm of the amplitude and the irradiance are related by (Goodman, 1985):

$$\chi = \frac{\left[\ln\left(\frac{I}{A^2}\right)\right]}{2}$$

(16)

According to the Rytov approximation, the variance of the logarithm of the amplitude $\langle\chi^2\rangle$ for a plane wave is (Goodman 1985):

$$\langle\chi^2\rangle = \sigma_\chi^2 = 0.307 C_n^2 L^{11/6} k^{7/6}$$

(17)

It has been shown that the above equation (13) is a good approximation for values of $\sigma_\chi^2 < 1$ (Wheelon, 2003]. The variance of the logarithm of the intensity is related to the variance of the logarithm of the amplitude of (Wheelon, 2002).

$$\sigma_{\ln I}^2 = \left\langle\left|\ln I - \langle\ln I\rangle\right|^2\right\rangle = 4\sigma_\chi^2$$

(18)

And

$$\sigma_{\ln I}^2 = 1.23 C_n^2 L^{11/6} k^{7/6} = \sigma_R^2$$

(19)

Where σ_R^2 is known as the Rytov variance. The Rytov variance for an infinite plane wave gives information about the strength of the fluctuations in the irradiance and hence gives us an idea of the strength of the atmospheric turbulence. Table II shows the relationship between values of Rytov variance and the strength of fluctuations (Wasiczko, 2004).

Table 4: Typical values of turbulence for turbulence levels from weak to strong

Strength levels of turbulence	Rytov variance
Weak	$\sigma_R^2 < 0.3$
Medium	$\sigma_R^2 \sim 1$
Strong	$\sigma_R^2 \gg 1$

Table 5: Models for irradiance distributions (*PDF: Probability destribution function)

Probability distribution function	Theory	Features	Application
Rician [Wheelon, 2001]	Born approximation	Little agreement with experimental data	Extremely weak turbulence regime
Lognormal [Tatarski, 1970]	Rytov approximation	Matching moments with experimental data	Weak turbulence regime
Negative Exponential [Andrews, 2005]	Heuristics	Easy to handle analytically	Saturation regime
I-K [Andrews, 2005]	Modulation effects of large scales to small scales	Difficult to relate PDF* parameters with the turbulence ones	Strong Turbulence
Lognormal – Rician [Andrews, 2005]	Modulation effects of large scales to small scales	Difficult to relate PDF* parameters with the turbulence ones	Strong Turbulence
Gamma-Gamma [Andrews, 2005]	Modulation effects of large scales to small scales	Its parameters are directly related to the turbulence.	Weak to strong turbulence

Another parameter used to compare the magnitude of the fluctuations of the irradiance is the transverse coherence length of an electromagnetic wave at optical frequencies (Wheelon, 2001). The coherence length for a plane wave is obtained from (Wheelon, 2003).

$$\rho_0 = \left(1.46k^2LC_n^2\right)^{-3/5}$$
(20)

For a spherical wave coherence length is given by (Wheelon, 2003)

$$\rho_0 = \left(0.546k^2LC_n^2\right)^{-3/5}$$
(21)

The coherence radius ρ_0 defined by Fried (Andrews, 2005) is:

$$r_0 = 2.099\rho_0$$
(22)

The meaning of ρ_0, can be interpreted as follows: the phase in the wave front does not experience fluctuations in the sense of mean square root of greater than one radian at a distance ρ_0 wavefront at the receiver (Wheelon, 2003). The following table summarizes and compares differents models for irradiance distribution that have been proposed by several authors (Andrews, 2005), (Zsu, 2002).

Phase Variations

The phase fluctuations not are usually take into account in incoherent optical wireless communication systems. However, in coherent optical wireless communication systems they should be considered. The phase fluctuations are caused by large eddies including those of outer scale (Goodman, 1985). It follows that the analysis of phase fluctuations are based on geometrical optics. Diffraction effects due to small-scale inhomogeneities have little effect on the result obtained based on geometrical optics (Wheelon, 2001). The complex phase disturbance [equation (40)], the phase $S(r,L)$ can be expressed (Tatarski, 1971) as:

$$S(\bar{r},L) = \frac{1}{2i}\left[\psi(\bar{r},L) - \psi^*(\bar{r},L)\right]$$

(23)

considering that the turbulence in the atmosphere is homogeneous and isotropic, the phase variance (Andrews, 2005) is :

$$\sigma_S^2 \cong 4\pi^2 k^2 L \int_0^\infty \kappa \Phi_n(\kappa) d\kappa$$

(24)

the phase covariance function or the spatial covariance function for plane wave can be expressed as:

$$B_{S,pl}(\rho,L) = 0.78 C_n^2 k^2 \left(\frac{\rho}{\kappa_0}\right)^{-5/6} K_{5/6}(\kappa_0\rho)$$

(25)

where K is the modified Bessel function of second class. The temporal covariance function can be obtained from the spatial function using the frozen turbulence hypothesis of Taylor (Zhu and Kahn, 2002) replacing $\rho = V\perp$ where $V\perp$ is the average wind speed transverse to the propagation path. Therefore, the spatial covariance function is (Wheelon, 2003].

$$B_{S,pl}(\tau,L) = 0.78 C_n^2 k^2 \kappa_0^{-5/3} (\kappa_0 V_\perp \tau)^{5/6} K_{5/6}(\kappa_0 V_\perp \tau)$$

(26)

The power spectrum of phase variations was first published in the work of (Clifford, 1970) and can be obtained using the Wiener Khintchine theorem (Tatarski, 1970) as shown below. Applying the Fourier transform of the function of temporal phase covariance, we obtain the temporal spectrum of phase variations [Tatarski, 1970].

$$S_{S,pl}(\omega) = 4\int_0^\infty B_{S,pl}(\tau,L)\cos(\omega\tau)d\tau$$
$$= 3.13 C_n^2 k^2 L \kappa_0^{-5/3} \int_0^\infty (\kappa_0 V_\perp \tau)^{5/6} K_{5/6}(\kappa_0 V_\perp \tau)\cos(\omega\tau)d\tau$$

(27)

Evaluating the integral gives [Wheelon, 2003] we obtain the aproximated expression

$$S_{S,pl}(\omega) = \frac{5.82 C_n^2 L V_\perp^{5/3}}{\left(\omega^2 + \kappa_0^2 V_\perp^2\right)^{4/3}}$$

(28)

Polarization Fluctuations

The electromagnetic field is characterized by an electric field and a magnetic field which are vector quantities. The direction taken by the electric field vector at each point along the path is defined by the polarization of the field (Fowles, 1968). There have been several studies to estimate the magnitude of the change of polarization in an optical frequency electromagnetic signal as it travels through the turbulent atmosphere (Collet, 1972) (Strohbehn, & Clifford, S. 1967). These studies conclude that the change in the state of polarization of a beam traveling in a line of sight path in the turbulent atmosphere is negligible. Depolarization is usually measured as the ratio between the average intensity of the orthogonal field component and the incident plane wave (Wheelon, 2003). Under certain considerations depolarization can be obtained through:

$$\delta \text{Pol} = 0.070 C_n^2 \left(\kappa_0 \right)^{7/3} k^{-2} \tag{29}$$

Various expressions have been obtained to determine the depolarization of an electromagnetic field at optical frequencies, considering quasi-monochromatic light sources and the results are similar. For example for $L = 1500$m, $\lambda = 1550$ nm and $C_n{}^2 = 1 \times 10^{-13}$ the depolarized component is 2.1×10^{-18} smaller in terms of the polarized component (Wheelon, 2001).

Arrival Angle Fluctuations

Fluctuations on the angle of arrival is another effect of atmospheric turbulence and seriously affects the performance of the communications system (Andrews, 2005). The movement of the centroid of the spot intensity on the receiver due to local inhomogeneities in the transmitter are responsible for this phenomenon. In the case of of non-coherent optical wireless communications wireless systems, this effect can be decreased by expanding the transmitted beam, so you always get intensity above the detection threshold to the receiver at the expense of the decrease in the average intensity (Wheelon, 2003). A more sophisticated technique is the use of pointing and tracking mechanisms of the centroid of the optical signal which makes adjustments on both the receiver and transmitter to ensure the highest possible alignment between them (Hemmati, 2006). Another way of reducing the effects of the variations on the angle of arrival is the use of adaptive optics, which correct these variations provided that the receiver aperture is large enough (Wheelon, 2001), (Andrews, 2005). The variance of the perturbations of the angle of arrival are obtained from the following equation (Wheelon, 2003).

$$\left\langle \delta \theta^2 \right\rangle = 2\pi R \int_0^\infty d\kappa \, \kappa^2 \Phi(\kappa) \tag{30}$$

Statistical Models of Wireless Optical Channel

As mentioned above, various probability distribution functions have been proposed to describe the statistical behavior of atmospheric optical communications channel. It was found that the amplitude distribution (or intensity) and phase is dependent on the theory of propagation of optical beams used. The phase distribution is obtained from geometrical optics and found that is suitable for the various regimes of turbulence (Andrews, 2005). Under the condition that the beam path is much larger than the size of the outer scale, based on the application of central limit theorem phase fluctuations of the optical signal is Gaussian and several experiments have supported the outcome (Clifford, 1970).

System Design

This section will show the basics for the design of a OWC link. The power budget of an optical link must consider different impairments that affect the system performance such as : a) finite transmission power, b) optical gains and losses, c) Receiver sensitivity, d) propagation losses, e) electronics noise, f) phase noise of optical sources g) imperfect synchronization for coherent detection optical carrier, among others. First, we determine the fade margin between the transmitted optical power and minimum receiver sensitivity needed to establish a specified BER. It also should be considered the system margin (M_s), to compensate for the degradation of components and temperature factors. It is required to estimate a margin of availability (M) or link power budget, which is given by the following equation.

$$M = L_f - L_{tur} - L_{prop} - L_{poin} - L_{atm} - M_s \qquad (31)$$

Where:

L_f: fade margin

L_{tur} : turbulence losses

L_{prop} : propagation losses

L_{poin} : Pointing losses

L_{atm} : atmospheric losses

M_s : system margin

Parameters to be considered in the design are: wavelength, transmission rate, signal to noise ratio (SNR), link distance, diameter of the optical transmitter and receiver antennas, transmitter power and receiver sensitivity. We describe below the relationship among the parameters mentioned.

Fade Margin

It is defined as the amount of the total losses allowed by the system to perform the optical link and is obtained from the equation:

$$L_f = P_{Tx} - P_{sens}$$

(32)

Propagation Losses

Propagation losses are given by (Santamaria A., Lopez-Hernandez F.J., 1994):

$$L_{prop} = 10\log_{10}\left(\frac{4\pi Z}{\lambda}\right)^2$$

(33)

where Z is the distance between the transmitter and receiver.

Turbulence Losses

These losses take into account the effects of the variation of intensity of the laser beam due to atmospheric turbulence (scintillation) and can be estimated through:

$$L_{turb} = 10\log_{10}\left[1+\left(\frac{\Omega_0}{\Omega_{turb}}\right)^2\right]$$

(34)

Where

$$\Omega_0 = \frac{2\lambda}{\pi D_{Lens\_Tx}}$$

(35)

With D_{Lens_Tx} is the lens transmitter diameter, and

$$\Omega_{turb} = \frac{\lambda}{\pi \rho_o}$$

(36)

where ρ_b is the coherence radius.

Pointing Losses

Pointing losses are due to misalignment between the transmitter and receiver the which causes reduction in the power captured by the receiver (A. Santamaria, FJ LopezHernandez, 1994), are given by (A. Santamaria, FJ Lopez-Hernandez, 1994)

$$L_{pointing} = 4.3229\left(\frac{\phi_e}{\Omega_0}\right)^2$$

(37)

Where φ_e is the boundary angle of diffraction-limited beam of the transmitter and is given by

$$\phi_e \cong \frac{\lambda}{2D_{Lens\_Tx}}$$
(38)

Atmospheric Losses

They appears when the particle causing the scattering has the diameter equal to or greater than the wavelength of the radiation signal. These lossess are due to atmospheric gases (Beer's law). The attenuation and scattering coefficients are related with the visibility (Kim et al).

Geometric Losses

Geometric path losses for a FSO link depends on the beamwidth of the optical transmitter (θ), the path length (L) and the receiver aperture area (D_r) (Figure 8):

$$L_{geo} = 20 \log\left(\frac{\theta L}{D_r}\right) \quad dB$$
(39)

L=transmitter-receiver distance

θ=Beam Divergence

D_r=Receiver diameter

Transmitting and Receiving Antenna Gain

The gain of the transmitting antenna for free space is given by (A. Santamaria, FJ LopezHernandez, 1994)

$$G_{Tx} = 10 \log_{10}\left(\frac{2}{\Omega_0}\right)^2$$
(40)

The receiving antenna gain is given by (A. Santamaria, FJ Lopez-Hernandez, 1994)

$$G_{Rx} = 10 \log_{10}\left(\frac{4\pi A_r}{\lambda^2}\right)$$
(41)

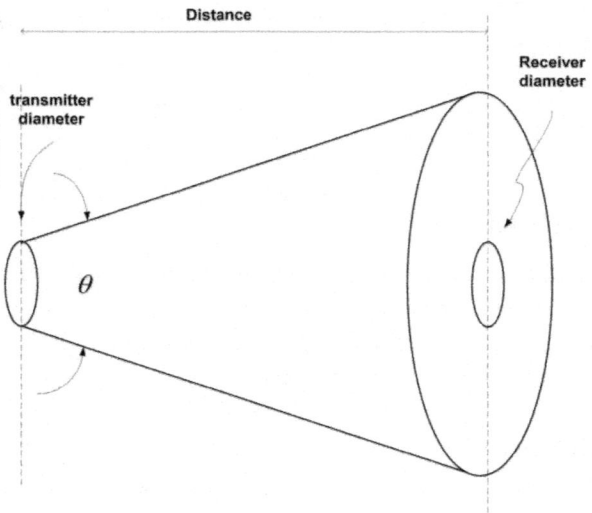

Figure 8: Geometric losses scheme.

MITIGATING THE EFFECTS OF TURBULENT OPTICAL CHANNEL

One of the problems to be resolved in optical communication systems is to reduce the effects of turbulence, i.e. the scintillation and variations of the angle of arrival of the beam. Various techniques are used to reduce these phenomena. Among them we can mention the use of encryption, the use of large aperture receivers, using alignment systems, spatial diversity and amplifiers using erbium-doped fiber (EDFA).

Using Coding to Reduce the Effects of Turbulence in OWC Systems

One way to improve the performance of wireless optical communication systems is the use of channel coding techniques. Several studies have been conducted to study the effect of the use of channel coding techniques in conditions of strong turbulence (Tisftsis, 2008) which is the scenario that offers the worst operating conditions. Pulse modulations such as PPM (Pulse Position Modulation) have been analyzed under the effects of weak turbulence (Hemmati, 2006). These results indicate the need for error correction in the receiver (FEC) to make communication possible under these conditions (Ohtsuki, 2003).

Large Aperture Receiver

It is known that for incoherent optical communications systems, such as IM-DD systems, the use of larger receiver apertures, increase the optical power collected leading to a reduction in scintillation. This effect is known as aperture averaging. This means that the larger the diameter of the receiving aperture, the power collected is higher, the signal has a better signal to scintillation noise ratio and the photo-current fluctuations are reduced (Fried, 1967).

Tracking and Pointing Systems

To reduce the effects of drift in the beam and the transmitter-receiver misalignment, phenomena that reduce the performance of wireless optical communication systems, mechanical systems can be used to correct both transmitter and receiver to compensate for variations tilt and pitch. This is possible because both changes occur at speeds of tenths of seconds (corresponding to frequencies below 100 Hz) (Andrews, 2005).

Use of Spatial Diversity to Mitigate the Effects of Turbulence

One way to reduce the effect of signal fading due to turbulence, which is mainly caused by beam wander, is the use of arrangements of receivers (Andrews, 2005).

Erbium-doped Fiber Amplifiers (EDFA)

The use of EDFA in the receiver avoids the use of high power transmission. It has been shown that the use of these devices also reduces the scintillation due to increased average received optical power (Franz & Jain, 2000), but these devices could be expensive for certain applications of OWC systems.

METHODS OF MODULATION AND CODING

Traditionally, wireless communications systems use optical modulation formats OOK (OnOff Keying), which is also widely used in fiber optic systems and is characterized by its simplicity and robustness. This system consists of intensity modulated optical carrier and digital information is sent with the presence or absences of the optical carrier. Other modulation techniques have been used in optical wireless communication systems, such as pulse position modulation and the use of phase-modulated subcarriers. One of the problems present in the transmission of optical signals is scintillation, which reduces the optical power available at the receiver for periods that can be several milliseconds to values below the detection threshold and thus interruption link. Different alternatives for the solution to this problem have been proposed and analyzed. You can increase the received optical power using erbium-doped fiber amplifier (EDFA). The atmospheric turbulence reduces the received optical power which is caused by the low frequency components of the scintillation and is expressed as the displacement of the centroid of the spot or footprint of the beam in the plane of the receiver (beam wander).

Incoherent Optical Communication Systems. OOK Modulation

Within the methods of direct detection and intensity modulation, one of the most used techniques is the On-Off Keying modulation. For this modulation has been found that the probability of error (Andrews, 2005) is:

$$P_e = \frac{1}{2}\int_0^\infty f_1(i)\,\mathrm{erfc}\!\left(\frac{\mathrm{SNR}(i)}{2\sqrt{2}}\right)di$$

(42)

where SNR is the signal to noise ratio as a function of intensity and erfc is the complementary error function. $f_1(.)$ is the probability density function of changes in signal strength.

Use of Subcarriers

Basically, the resurgence of practical OWC systems is due to the technological developments of the systems of fiber optic communications. One of the techniques used to improve the performance of OWC systems is the use of subcarriers. In this method, the laser beam intensity is modulated by an electrical signal derived from a combination of these subcarriers. Figure 9 shows the block diagram for subcarrier intensity modulations systems.

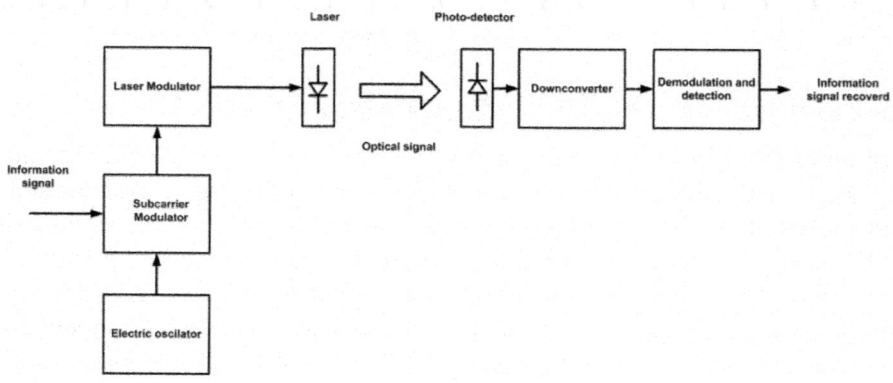

Figure 9: Subcarrier intensity modulation OWC.

Coherent Optical Communication Systems

As indicated above, the current optical communications systems are based on incoherent modulation techniques which are relatively simple to implement and robust, but its theoretical performance is below the coherent modulation format. This type of system has advantages in relation to sensitivity, frequency selectivity and increased lodging capacity of channels in the bandwidth of the optical carrier. The coherent optical communication systems in atmospheric space applications have interesting characteristics that make them attractive for potential commercial use. For example, the homodyne detection of binary phase modulated signals (BPSK), the quantum limited is obtained with only 9 photons per bit, when in the OOK systems are needed 38 photons per bit. The BER for BPSK modulation is an average over the all possible intensity levels of a given probability density function, $p_I(I)$ without regard phase noise (Sánchez, 2008):

$$BER = \int_0^\infty p_I(\xi) \text{erfc}\left(\sqrt{SNR(\xi)}\right) d\xi \tag{43}$$

The optical phase synchronization, and control of the state of polarization are the main challenges for the practical implementation of coherent systems using optical fiber as transmission medium (Kazovsky, 2006).

In the case of wireless systems, in clear sky conditions, the state of polarization suffers little variation and these changes are slow (Hodara, 1966) (Wheelon, 2001), but it is required that the state of polarization of the signal optical input matches the local oscillator. Carrier synchonization is neccesary to achieve the demodulation in coherent systems. The phase modulation techniques are usually suppressed carrier. Techniques such as injection locking, optical phase lock loop (OPLL) can not be used directly to lock the local oscillator phase [Kazovsky, 1986]. With the advent of high-speed digital components, the compensation of polarization, as well as other phenomena of the optical channel can be obtained in the electrical domain, opening up new possibilities for the practical implementation of optical communication systems consistent (Sánchez, 2008), (Arvizu, 2010). Figure 10 shows the block diagram of a coherent optical wireless communication system which shows the possible subsystems required to enable proper operation. At the transmitter, the optical phase modulation is performed while at the receiver is used an phase lock loop (PLL) to maintain synchronized the optical carrier signal with the optical local oscillator [Kazovski et al, 1995], and a state of polarization control system (Sánchez 2008), (Arvizu, 2010), as well as a balanced photo-reception stage.

Due to the loss of spatial coherence cannot use aperture averaging in coherent optical communications systems and diameter in the aperture receiver must be smaller than the coherence distance r_0 (equation 22). For example, with L=1500 m, λ=1.550μm and $C_n^2 = 7\times10^{-13}$, r_0=2.5 cm (Figure 11).

However, the small apertures require the use of less divergence beams so that more optical power is collected by the receiver, which involves the use of pointing and tracking systems more fine and precise, making the system more complex and expensive. Another solution is the use of spatial diversity system. The space diversity coherent systems require that each unit receiving signals are processed individually before combining it and then perform the symbol detection process (Arvizu et al, 2010). That is, it requires that the signals to be synchronized in phase due to the loss of spatial coherence so that the combination of signals is not destructive and attenuates the signal received. This process can be performed optically or electronically (Arvizu et al, 2010). The distance between these coherent diversity receivers must be greater than r_0, so that the signals collected by each unit are uncorrelated. Other proposed systems is the use of OWC systems with spatial diversity and coherent detection using (linear post detection combiner) (PDLC), which uses "n" receivers and develops individually detection by estimating the symbol for each (hypothesis "1" and "0") then becomes the weighting of the signal with better signal to noise ratio which is selected to obtain the output data (Arvizu et al, 2010).

Coherent optical communications systems offer several advantage in deep space applications, such as high sensivity, which is important because of the small signals existent in this scenery and the absence of atmospheric turbulence. Additionally coherent receivers have an inherent frequencial selectivity, as well as rejection of the background radiation, characteristics important in deep space applications. Next generations of optical wireless communications could use differents strategies for reduce the turbulence effects. Adaptive optics is a technology utilized for improve the performance of astronomical telescopes by reducing the wavefront distortions and can be used in OWC systems. However, still is a technology expensive for terrestrial OWC applications.

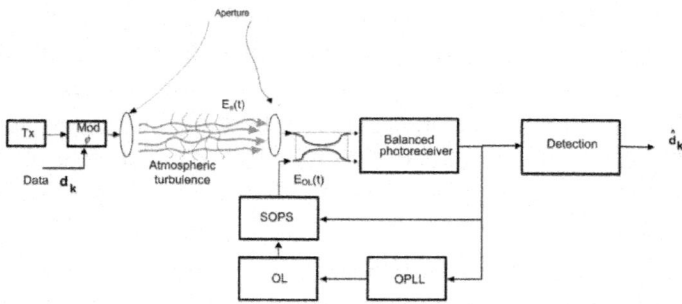

Figure 10: Block diagram of the Coherent optical wireless communications system. SOPS: State of polarization system; OL: Local Oscillator: OPLL: Optical Phase lock loop.

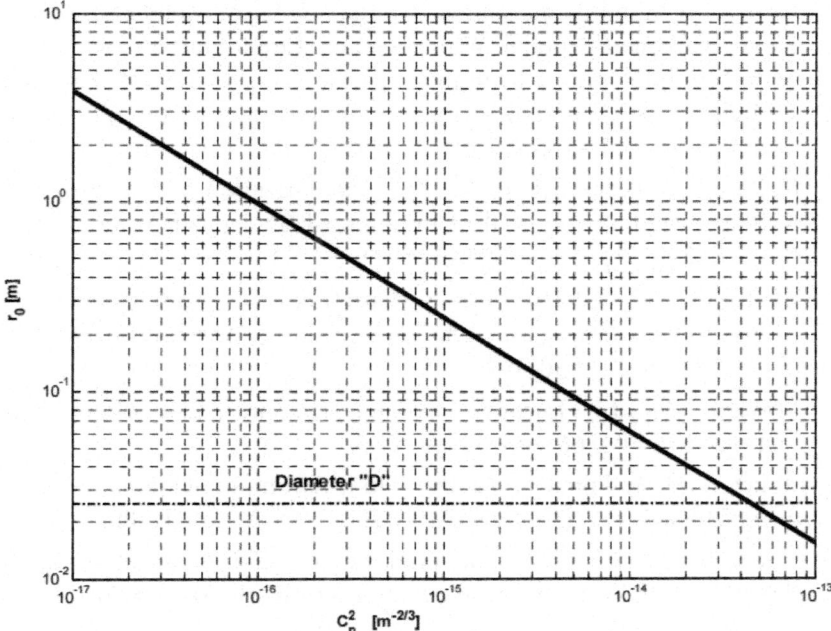

Figure 11: Coherence diameter as function of the refractive index structure constant.

CONCLUSIONS

In this chapter, the wireless optical communication systems have been discussed from first principles to systems that use different techniques to improve their performance. Different atmospheric channel characteristics have been emphasized and in general have shown the most relevant such as scintillation, the variations of the angle of arrival, the attenuation due to atmospheric gases and the effects of weather conditions. We analyzed the performance of communications systems for detecting incoherent modulations (OOK) and coherent (BPSK).This technology is becoming commonly used in civil applications and in the future be developed to have a scope similar to fiber optic systems in scope and availability.

REFERENCES

1. Agrawal, P, Govind (2005). Light Wave Telecommunications Systems John Wiley and Sons, Inc, ISBN -13 978-0-471-21572-2, New York.

2. Andrews, L. C. & Phillps, R.L. (2005). Laser beam propagation through random Media. SPIE Press, ISBN 0-8194-5948-8 Bellingham, Washington.

3.   Arvizu M. Mondragón, Sánchez L. Juan de Dios, Mendieta J. Francisco Javier, Coherent Optical Wireless Link Employing Phase Estimation with Multiple Beam, Multiple Aperture, for Increased Tolerance to Turbulence, IEICE Transactions communications, Vol. E93-B, No. 1, (January 2010), 226-229, ISSN 1745-1345.

4.   Collet, E. & Alferness, R. (1972). Depolarization of a laser beam in a turbulent medium. Journal of the Optical Society of America. Vol. 62, No. 4, (529-533), ISSN 0030-3941.

5.   Clifford, S.F. (1970). Phase Variations in Atmospheric Optical Propagation, Journal of the Optical Society of America. Vol. 61, No. 10, (529-533), ISSN 0030-3941.

6.   Fowles G. R. 1975. Introduction to modern optics. Dover Publications. 2 edition, ISBN 0486659577, New York.

7.   Franz, J. H. & Jain,V.K. (2000). Optical communications, components and systems. CRC Press. ISBN 0-8493-0935-2, New Delhi.

8.   Fried. D.L. 1967. Optical heterodyne detection of an atmospherically distorted signal wave front. Proceedings of IEEE. Vol. 55, No. 1 (47-67).

9.   Gagliardi, Robert M. and Karp, Sherman (1995). Optical Communications, John Wiley and Sons, Inc., Second Edition, ISBN 978-0471542872, New York.

10. Goodman, J. W. (1985). Statistical Optics. Wiley and Sons. ISBN 0471015024, New York.

11. Hemmati, H. (2006). Deep Space Optical Communications, John Wiley and Sons, Inc., ISBN 978- 0-04002-7.

12. Kazovsky, L., Benedetto, S., Willner, A (1996). Light Wave Telecommunications Systems. Artech House, Inc. Norwood, ISBN 0-89006-756-2.

13. Kedar, D. y Arnom, S. (2003). Optical wireless communications through fog in the presence of pointing errors. Applied Optics. Vol. 42 No.24, (4946-4954) ISSN 2155-3165.

14. Kim, I.I, Mc Arthur, B., Korevaar, E. (2000), Comparison of Laser Beam Propagation at 785 nm and 1550 nm, in Fog and Haze for Optical Wireless Communications. Proceedings of SPIE Optical Wireless Communications III, Vol. 4214, (26-37)

15. Lambert, G. Stephen and Casey, L. William. (1995). Laser Communication in Space. Artech House, inc., Norwood. ISBN 0-89006-722-8.

16. Ohtsuki, T. (2003). Performance analysis of atmosferic optical PPM CDMA systems. Journal of Lightwave Technology. Vol. 21 No. 2 (406-411), ISSN: 0733-8724.

17. Osche, G. 2002. Optical detection theory for laser applications. John Wiley and Sons, ISBN 0-471- 22411-1, New Jersey.

18. Qingchong L., Chunming Q., Mitchell G., & Stanton S. (2005). Optical wireless communications networks for first-and last-mile broadband access. Journal of Optical Networking. Vol. 4 No. 12 (807-828), ISSN 1536-5379.

19. Santamaria A., López-Hernández F.J. (1994). Wireless LAN systems, Artech House, ISBN 9780890066096.

20. Sánchez L. Juan de Dios, Arvizu M., Arturo, Mendieta J., Francisco (2008). Optical Phase Estimation in an Urban Wireless Communications, IEICE Transactions communications, Vol. E91-B, No. 7, (July 2008), 2447-2450, ISSN 1745-1345.

21. Strohbehn, J.W., y Clifford, S. (1967). Polarization and angle-of-arrival fluctuations for a plane wave propagated through a turbulent medium. IEEE Transactions on Antennas and Propagation, Vol. 15, No. 3 (416-421), ISSN 0018-926X.

22. Tatarski, V.I. (1971). The effects of turbulence atmosphere on wave propagation. The National Oceanic and Atmospheric Administration. U.S. Departament of Commerce, ISBN 07065-0680-4, Springfield, VA.

23. Tisftsis T.A. 2008. Perfomance of heterodyne wireless optical communications systems over gamma atmospheric channels. Electronics letters. Vol. 44, No. 5, (373-375), ISSN: 0013-5194

24. Wasiczko, L.M. (2006 Techniques to mitigate the effects of atmospheric turbulence on free space optical communications link. Ph.D Thesis. University of Maryland. 135 pp.

25. Willebrand, H. Ghuman, B.S. (2000). Free-space optics: enabling optical connectivity in today's networks", Sams Publishing, Indiana, ISBN 0-672-32248-x, USA.

26. Wheelon , A. D. (2001). Electromagnetic scintillation Vol, I. Geometrical optics. Cambridge University Press. Cambridge, ISBN 0521-801982

27. Wheelon, A.D. (2003). Electromagnetic scintillation Vol, II. Weak Scattering. Cambridge University Press. Cambridge, ISBN: 0521801990.

28. Zhu, X., Kahn J. M. (2002) Free Space Optical Communications Through Atmospheric Turbulence Channels, IEEE Transactions on Communications, Vol. 50, No. 8, August 2002. ISSN 0090-6778.

Chapter 9

CONNECTIVITY PREDICTION IN MOBILE VEHICULAR ENVIRONMENTS BACKED BY DIGITAL MAPS

Robert Nagel and Stefan Morscher

Institute of Communication Networks, Technische Universität München Germany

INTRODUCTION

As mobile ad-hoc networks gain momentum and are actively being deployed, providing users and customers with ubiquitous connectivity and novel applications, some challenges implied especially by the mobility of users have not yet been solved. Generally, it can be stated that modern applications impose higher requirements on the underlying communication solutions: more bandwidth, less packet loss, less delay and more reliability of services in terms of availability. These performance metrics are commonly termed Quality of Service (QoS). Due to the variability of node locations in mobile networks, the experienced QoS is highly time-variant. We have discussed in Nagel (2010a) that the level of attained QoS ultimately results from a proper combination of connectivity, i.e., the communication relations in a network, the chosen (and usually invariant) medium access (MAC) protocol and the traffic that is injected into the network at the nodes. If a certain level QoS is desired in a mobile wireless network, at least one of these three properties has to be actively controlled.

We have demonstrated that through controlling the amount of traffic that is injected by the nodes, effective distributed mechanisms can be employed that are, given minimal information about nodes' connectivity, able to provide (and even guarantee) a certain level of QoS. These mechanisms, however, are based on the current connectivity of the network and are effective only at present time. Should an application require a certain amount of QoS over a larger period of time, additional provisions become necessary. Although it is possible to control connectivity in certain boundaries (for instance through power control or adaptive antennas) and at a certain cost, the fundamental physical causes of connectivity themselves (location, mobility, and wireless channel state) cannot be influenced by the application as they are dictated by the user's behavior and the environment. It is, however, possible to anticipate a network's future connectivity – at least for a certain time horizon – and to compute the resulting future QoS. Upon this information, applications, services, and routing protocols could be parameterized accordingly: as an example, if the future QoS of a connection using a certain route is predicted to fall below a necessary level due to a link break, the expected remaining time until the link actually breaks could be used to proactively find and set up a backup route that uses other, potentially more stable links. Also, if a connection was to be set up for a limited time, it may be very helpful to assess if the required QoS can actually be provided by the network for the desired duration before the connection is actually established. While other work mainly uses mobility prediction in cellular scenarios to estimate hand-over times, or to support ad-hoc routing in random-mobility ad hoc scenarios, this chapterfocuses on connectivity prediction in the special case of vehicular networks.

Networked vehicular nodes can be assumed to adhere to certain rules that constitute drivers' basic behavior: they move along roads and try to avoid collisions with obstacles, such as buildings and other cars. Founded on the vehicular scenario constraint, we present an algorithm that predicts the location of vehicles based on their current state (position and velocity) and information from digital street maps obtained through the Open Street Map (OSM) project. A filter-based, self-adaptive velocity prediction algorithm is used to model the user-inflicted velocity changes. Using their current positions and the predicted velocities, possible future positions of cars on the street grid and their respective probabilities can be determined. Although the main focus of this chapter is on mobility prediction, we discuss an effective channel parameter estimation technique and propose to predict the network's future connectivity using an adaptive channel model. It should be noted that the proposed position prediction mechanism does not completely exhaust all opportunities provided by the vehicular scenario. For instance, we assume that vehicles have no information about other vehicles' missions, i.e., the planned route through the road grid. Furthermore, we make no assumptions about other vehicles' capabilities (in terms of maximum acceleration and deceleration, yaw rate, etc.). Also, we do not consider environmental properties, such as weather, street and traffic conditions, etc.

We will, however, point out and discuss the potential spots where these additional informations could be exploited to further augment the proposed algorithm. In the following Section, we present an overview of current work. Subsequently, in Section 3 we formulate the problem mathematically and in Section 4, we describe our algorithm that predicts the future positions of vehicles according to their actual state and a complemental digital road map. In Section 5, we will discuss why the channel model presented in Equation 1 is not sufficient under all conditions and present methods that can adapt to the environment. In Section 6, we discuss some simulation results. Section 7 summarizes this chapter and gives an outlook on further work.

RELATED WORK

One possible way of predicting a network's future connectivity is to use a model that reflects the individual mobility properties of a node. Given the knowledge of the initial position velocity of a node, a future position could be projected by multiplying the velocity vector with the desired time interval. Obviously, this approach does not account for changes in the length and/or direction of the velocity vector. Several more sophisticated approaches have been suggested and today, mobility prediction has become a common research topic in wireless networks.

Due to the distinct characteristics of vehicular ad-hoc networks, especially the high speeds and restricted degree of freedom in the movement of vehicles, most of the work on prediction for ad hoc networks is too general and thus inappropriate for vehicular networks. Nevertheless, some approaches are discussed here because they give an overview of mobility prediction in general. Material specific to mobility prediction in vehicular networks is very rare and the topic is often neglected in works on vehicular ad hoc networks.

Kaaniche & Kamoun (2010) presents an approach for mobility prediction using neural networks. Although it is not specifically designed for vehicular networks it should perform better than other general approaches as it is independent of the underlying mobility model. A trajectory is calculated for multiple steps in the future using several past positions in quite a similar manner as the adapting FIR filter for velocity prediction presented in this work.

However, the approach does not use any map material and hence the predicted positions may lie far off the road and may thus be unrealistic. The approach using neural networks could be used for velocity prediction in the constellation presented in this works to substitute the FIR filter, however it is expected to perform in a very similar manner and the FIR filter seems less complex to implement.

A similar approach for mobility prediction using spatial contextual maps and Dempster-Shafer's theory for decision making is formulated in Samaan & Karmouch (2005). A framework is presented that allows prediction of the users mobility trajectory based on various bits of contextual information from e.g. user profile and map data. The approach is motivated by the fact that contextual information is becoming more common for adapting services towards the users needs and it uses the additional information in order to predict the users mobility. The concept seems feasible for e.g. cell phone users traveling on foot but does not seem appropriate for vehicular networks as the only contextual information possibly available and relevant to the future mobility is the chosen route to the destination. A complex theory to combine evidence into a prediction is not necessary in this case.

Huang et al. (2008) suggests a prediction algorithm based on fuzzy logic that aims at the prediction of a possible link break or a congested link which then triggers the construction of an alternate route. Similar to our algorithm, the prediction of a link break is based on the prediction of the future vehicle speed, the basis on which the predicted distance to the vehicle can be determined. This requires the generation of a fuzzy rule base that is then dynamically trained using Particle Swarm optimization (which in our approach is done using the adaptive filter for speed prediction). The authors use similar ideas in terms of the speed prediction but implements a fundamentally different concept. Furthermore, it is focussed on route break prediction and hence the performance of the isolated velocity prediction compared to our algorithm cannot be easily evaluated. In Boukerche et al. (2009), the authors present some general thoughts on mobility prediction in vehicular networks and propose a simple prediction algorithm based on movement vectors in order to reduce the frequency of location beacons without introducing a higher mean error in respect to the positions used for routing packets.

In Rezende et al. (2009), the same authors introduce the Network Neighbor Prediction protocol (NNP) that uses the results from their prior works to predict new routes that are going to be available in the near future and to calculate the lifetime of those routes that are currently in use. These works show in their simulation results that mobility prediction is a useful and necessary aspect in vehicular networks and should be researched in greater detail than it currently is. Another approach, although developed in the context of a different problem, is described by Althoff et al. (2010). The authors compute the set of points that could be reached by vehicles within the prediction times, given the capabilities (minimum and maximum acceleration, yaw rate, etc). of the considered vehicles. The approach is computationally complex and requires a lot of contextual information.

Using the predicted position it should also be possible to predict the future connectivity to a certain extent using an appropriate channel model. In the context of vehicle-to-vehicle (V2V) communications, there is not yet a widely accepted channel model Paier et al. (2009). A common approach for characterizing a channel is to work out a theoretical channel model and then validate it against some appropriate measurements. Channel models are usually classified into stochastic and deterministic channel models, where deterministic channel models use ray tracing and similar techniques based on topological information about the environment in order to solve the the multi-path components (MPC) and derive a precisechannel characterization for a specific realization. Stochastic channel models, on the contrary, try to depict the statistics of the propagation channel in a more general sense that is not so much focussed on a particular situation. An intermittent approach is taken by geometry based channel models (as presented in Cheng et al. (2009)) that do use ray tracing; however, instead of using realistic modeling the calculations are based upon randomly placed objects.

In order to characterize a channel a number of parameters are used:

- The path loss exponent (PLE) α characterizes the average attenuation of the received signal.
- Large scale fading on the one hand refers to slow variations of received power due to shadowing by obstructing objects.
- Small scale fading on the other hand is caused by interference of different MPCs that result in fast fluctuations of the received power. Because these fluctuations are very hard to describe deterministically, they are usually described by means of statistics - most commonly by a Rayleigh distribution. • In order to determine how much power is carried by the respective MPCs a power delay profile (PDP) is used. The spreading of the received pulse in the time division - often referred to as the channels delay dispersion - is best described in a statistical way by the root mean squared delay spread.
- Because MPCs travel on different paths they experience different Doppler shifts. The root mean square Doppler spread describes the resulting spectrum widening of the received pulse and thus the frequency dispersion.

A large amount of research has been dedicated to the wireless channel in cellular networks. However, looking at the specifics of a vehicular channel, especially in the V2V case it soon becomes clear that its characteristics differ significantly from those of a cellular channel. On the one hand, antennas of both sender and receiver are mounted close to the ground in V2V communications, where with cellular systems usually one of them is mounted high above. This tremendously influences the propagation path of the signals and thus the channel characteristics in terms of diffraction and reflection. On the other hand, communications between vehicles commonly use the 5.9 GHz band which behaves significantly different than the 700-2100 MHz signals used in cellular systems in terms of attenuation and diffraction. Most importantly though, sender and receiver are moving at relatively high speeds in V2V scenarios, which invalidates the assumption of stationarity of the channel characteristics that is commonplace in channel models of cellular systems. That refers not only to a changing impulse response but also to a change of its statistical properties (fading distribution, PDP and Doppler spectrum) Molisch et al. (2009). According to Maurer et al. (2004), Doppler shift and Doppler spread characterize the time-variant behavior of the V2V channel mostly due to movement of the communicating vehicles and the adjacent vehicles.

This section highlights and discusses some of the works into vehicular channel modeling in the context of connectivity prediction - a topic that has not yet received much attention in literature. In Matolak et al. (2006), the authors describe a statistical V2V channel model that is restricted to small scale fading. It uses a tapped delay line model, each tap representing a multi path components received with a certain delay. Each tap has an on/off switching process modeled by a first order Markov chain allowing for persistence parameterization. In general, taps with longer delays have less probability of being on due to their lower energy. Tap amplitudes are modeled using the Weibull distribution where different parameters are proposed for different taps, based on some measurements. The authors differentiate betweendifferent scenarios, in some of which the Weibull parameters are "worse than Rayleigh" ($\beta < 2$), a phenomenon that is often called severe fading.

Maurer et al. present a geometry based IVC channel model in Maurer et al. (2004). They first try to model the dynamic road traffic and the environment adjacent to the road and then try to evaluate multi-path wave propagation through means of ray tracing. The road traffic model is based on the so called Wiedemann model and uses results from the authors previous works. As it seems very difficult to obtain real data with the necessary level of detail and the coverage, a stochastic model is utilized in order to place objects in the surroundings of the road. Different morphographic classes are defined for urban, suburban and highway scenarios that are assigned specific probabilities for different types of objects (trees, buildings, cars, bridges, traffic signs, etc.). Multi-path components are represented by rays, each of which can experience several propagation phenomena like diffraction or reflection. By calculating consecutive snapshots, a time-series of channel impulse responses can be obtained that classifies the channel for the current surrounding. The authors present measurements that validate the channel model with a standard deviation of less than 3 dB in both line-of-sight (LOS) and non-line-of-sight (NLOS) scenarios.

Paier et al. (2009) presents some measurements of V2V propagation in suburban driving conditions using GPS receivers. The authors on the one hand derive both a single slope and a dual slope path loss model from their results where the better dual slope model achieves deviations between 2.6 and 5.6 dB compared to the measured path loss. However, they find that received power is significantly less if no LOS propagation is possible. Fading on the other hand is modeled using a Nakagami distribution with variable parameters as already proposed in other works. While the distribution is Rician $\beta > 2$ as long as a LOS component is present, it turns out that fading can be "worse than Rayleigh" $\beta < 2$ once the LOS connection is lost intermittently at large distances between transmitter and receiver. Furthermore, the authors propose that the Doppler spread is dependent on the effective speed and the distance between transmitter and receiver. The dependance on distance is explained by the increasing number of scatterers at larger distances. Using this dependence, the authors present the speed-separation diagram that can help predict the expected Doppler spread and thus small scale fading characteristics at a certain distance.

In Molisch et al. (2009), the authors provide a survey on V2V channel models and measurements based on a variety of previous works on the subjects, some of which have already been discussed here. We recommend this paper as an introductory reading on the subject as it introduces important factors for channel characterization and includes a table that summarizes important parameters gathered from multiple measurement campaigns. Important aspects like environment characterization and antenna placements are also discussed that we omit here. One important result from the evaluated measurements is that at least path loss coefficients in V2V communication channels are rather similar to well-known cellular systems as long as a LOS connection is given. In terms of small-scale fading and Doppler spread, the results go alongside those presented in Paier et al. (2009).

The authors finally conclude that the amount of comparable measurements carried out on V2V channels is too insignificant in order to allow the formulation of a channel model that resembles the real-world V2V channel and important aspects such as antenna placement and shadowing by adjacent vehicles have not yet been sufficiently explored. Following this conclusion, an adequate prediction of channel quality seems challenging. Analogous to position prediction, an estimation of channel quality can be seen as a tradeoff between computational complexity and prediction accuracy. An approach involvingray-tracing similar to the one presented in Matolak et al. (2006) on the one hand produces rather adequate results if provided with the necessary extent of details concerning the surrounding environment (including moving and parked vehicles), building geometries, plants and road signs. However, it seems unrealistic and infeasible to supply an on-board connectivity prediction engine with this amount of knowledge. Measurements suggest a dual-slope model for the path loss exponent as a very simple approach. Small-scale fading is usually modeled using statistical models with strong dependency upon separation distance which limits the possibilities of a prediction to a qualitative worst case approximation. Paier et al. (2009) also identifies significant differences between LOS and NLOS cases in both path loss and fading statistics.

A sophisticated approach to predict the path loss exponent using a particle filter has been proposed in Rodas & Cascon (2010), based on a log-normal fading channel model in wireless sensor networks. Particles are initialized in a random state with their respective weights being iteratively updated to provide an estimation of the path loss exponent. Weak particles with low weights are periodically replaced to avoid degeneration. The filter is parameterized with the type of the fading distribution and its variance. The authors, too, show that the PLE changes significantly as soon as the LOS is lost.

PROBLEM STATEMENT

In Nagel (2010b), we have outlined how QoS provisioning based on a network's connectivity can be attained. The basis for the computation is the connectivity matrix $\underset{\sim}{C}$ that describes the communication relations between n networked nodes. Let $\chi(x_i, x_j)$ denote the channel function, taking as parameters the physical positions x of two vehicles in the environment. A very basic channel function could then read:

$$\chi(\underline{x}_i, \underline{x}_j) = \begin{cases} 1 & \text{if } \left\| \underline{x}_i - \underline{x}_j \right\| \leq r \\ 0 & \text{otherwise} \end{cases}$$

$$(1)$$

This means that two vehicles i and j are connected if they are located closer than the radio range r; if they are located further apart, they are not connected. The connectivity matrix $\underset{\sim}{C}$ is then defined as:

$$\underset{\sim}{C} = (c_{ij}), \quad c_{ij} = \chi(\underline{x}_i, \underline{x}_j)$$

$$(2)$$

Every node i is allowed to inject (source) traffic amounting to s_i into the network. Multiplying the source vector $\underline{s}$ with the connectivity matrix results in the load vector l:

$$\underline{l} = \underline{C}\,(\underline{1} + \underline{s}) \tag{3}$$

We have shown that the QoS criterion is fulfilled if the injected traffic is dimensioned so that each entry in the load vector l_i does not exceed a certain pre-defined threshold. For more detail, especially on the distributed algorithm, the reader be referred to the original paper. The problem with this approach, however, is that $\underline{s}|_{t_0}$ is only valid for the current connectivity matrix $\underline{C}|_{t_0}$. As it is desirable to fulfill the QoS criterion over a certain time Δt, we first need to predict the future physical positions of the vehicles, estimate the channel function and then deduce the prospective future connectivity matrix:

$$\underline{C}|_{t_0+\Delta t} = \left(\underbrace{\chi|_{t_0+\Delta t}}_{\substack{\text{Channel}\\\text{Estimation}}} \underbrace{\left(\underline{x}_i|_{t_0+\Delta t},\underline{x}_j|_{t_0+\Delta t}\right)}_{\substack{\text{Mobility}\\\text{Prediction}}} \right) \tag{4}$$

After that, the future source vector can be computed (Equation 3) and a decision can be made whether the current demand can be satisfied under the future network conditions and consequently, adequate measures can be taken.

MOBILITY PREDICTION

Generally, the spatial behavior of a vehicle is defined by two factors: On the one hand, speed and direction are controlled by the driver who adapts to the environment and the current situation. On the other hand, movement of a car is restricted to roads so the surrounding road topology is the major limiting factor. This is the key criterion that simplifies location prediction for vehicles compared to regular mobile users. Cars are usually not allowed to travel anywhere, they are bound to a relatively small portion of the world, the lanes. Combined with a small memory of past positions, the current velocity and direction of movement can be calculated. This further limits the amount of available future positions, as cars are usually not expected to u-turn spontaneously and velocity changes are bounded by the maximum deceleration and the maximum acceleration.

Concept

The prerequisite for the prediction is knowledge about a vehicle's current position, direction of movement and the surrounding road topology. The latter is provided by digital street maps (available, for instance, through the OpenStreetMap Project). All of these factors are very stable in terms of prediction. The destination or rather the mission of the car is assumed to be unknown to the algorithm, so at a crossroads basically all directions seem equally probable. The velocity of a car, however, is far less stable and predictable as it is directly controlled by the user and indirectly influenced by environmental factors such as traffic density, road signs and the weather. Especially abrupt speed changes are almost impossible to predict as they are often unexpected, even to the driver himself. The algorithm is sketched in Figure 1.

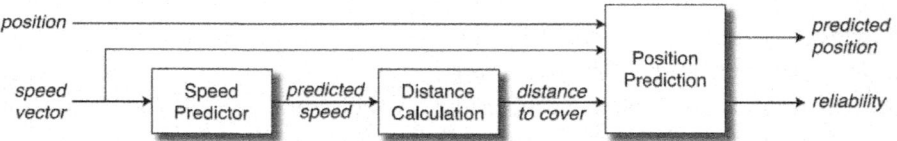

Figure 1: Algorithm Outline.

For speed prediction, we use a filter based approach that employs concepts of adaptive filters initially developed to adapt to varying channel conditions in wireless communications. Like the channel characteristics change depending on the environment, the speed change behavior of a car - or rather its driver - adapts to various environmental factors. This includes urban scenarios with steep velocity slopes and rural roads with fairly constant speeds. The character of the driver and the performance of the car also influence the prediction to a certain extent and are automatically taken into account by the adaptive filter. A self-adapting finite impulse response (FIR) filter approach based on a least-mean-squares (LMS) algorithm with relatively low depth seems ideal to adapt to both the personal behavior of a driver and the current situation. Using past and current velocities, an ideal weight vector for the past situation is calculated. Due to the low depth of the filter, the weight vector is rather unstable and consequently, it is combined with both the mean weight vector over the last iterations and a "boost" vector to improve reactivity at steep slopes. The resulting weight vector is then used to predict the future velocity, which is in turn used to calculate the distance covered in the desired interval.

The distance to cover, together with the current position and direction of movement, forms the input for the position predictor that outputs the predicted future location of the vehicle. In some cases, multiple positions are possible, for instance due to a crossroads between the current and the future position. In that case, the position that seems most probable to the algorithm is used as an output; however, internally a list of all possible locations is generated. In many situations, predominantly with cars traveling in sparsely populated areas or on highways, the prediction is rather reliable. In urban areas prediction reliability is reduced by intersections where a sudden change of direction can occur and a certain amount of past predictions may be invalidated. To make applications aware of such differences, an additional output variable was added to resemble the estimated reliability of the output.

Input Data

The algorithm requires a number of input data:

Position Data

Obviously, the algorithm requires knowledge about the actual position of a vehicle and a timestamp. The position data used in the performance analysis has been downloaded from the "GPS Tracks" section of the OpenStreetMap online portal. Selected tracks were chosen that were provided by users around the globe and thus constitute a rather broad basis of real life data. Additionally, own traces have been used. The temporal resolution of the recorded tracks was or has been resampled to one second. A statement about the spatial resolution is not generally possible as different positioning hardware from various vendors has been used for the sample data. However, we shall assume a positioning accuracy of a few meters.

Map Data

Also, the algorithm needs to be provided with map data of the area surrounding the actual position. This data, too, is provided by and downloaded from the open source OpenStreetMap project. It basically consists of an array of so-called nodes that are uniquely identified and reference a GPS position by latitude and longitude. A street is constructed by a list of subsequent nodes, forming a polyline that represents the shape of the street. Actual contiguous roads may be split apart, for instance if the name of a street changes or if two streets merge, on intersections etc.

Number of Steps to Predict

The major parameter influencing the algorithm. It is common in most parts of the algorithm and hence introduced in the high level diagram. Many parts of the algorithm also refer to it as n. Depending on the input data, the usual assumption is that one timestep equals one second. Most of the evaluations were done using a medium interval of prediction of 8 seconds - however results using different values are discussed in section 6.4.

Speed Predictor

A car typically moves in different classes of environments: urban, suburban, peripheral and highway. Each of those has different characteristics concerning the speed change of a car. On a highway speed changes are rare but usually with rather steep slopes whereas in urban areas, the speed is hardly ever constant for more than a few seconds. This allows for two different approaches in implementing a speed prediction algorithm. On the one hand, specialized algorithms could be engineered for all of the above scenarios and another algorithm that determines the algorithm that is most appropriate in an actual situation. In typical situations one would expect such an approach to give very accurate results, but clearly there are many situations where none of the implementations will be adequate.

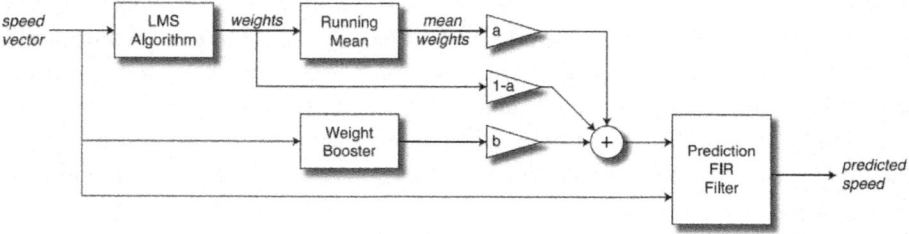

Figure 2: Speed Predictor Structure.

Furthermore, this approach involves increased efforts in development because multiple algorithms need to be designed and there is a high number of factors influencing the situation that are hard to quantize. On the other hand, it seems more appropriate to design an algorithm that automatically adapts to changing situations and as such can also adapt to factors like driver attitude and others mentioned above. This introduces some delay caused by the responsiveness of the adaption algorithm, but works also in an environment that cannot be properly classified into one of the above scenarios. In some situations, especially with quickly varying conditions, this may result in weaker performance than the approach discussed above, but the overall performance is expected to be better with less development efforts. A solution for this approach is discussed in the next sections.

Structure

The signal flow graph of the speed prediction is shown in Figure 2. The input variables are the current speed of the car and the number of steps n to predict. The only output is the predicted speed for the given time frame.

Prediction FIR Filter

The actual prediction is done in an FIR filter on the right hand side of the signal flow plan. It uses the current speed and a weight vector to predict the velocity from the last speed values. The length of the weight vector is given by the depth of the FIR filter - in the evaluations performed here a depth of 12 was used.

LMS Algorithm

The most important building block. It is the adaptive part of the algorithm and calculates an ideal weight vector from its two input values - the current velocity forms the desired signal, a delayed version forms the input for the algorithm. The weight vector is adapted with a fixed step size in the direction of steepest descent in order to achieve the minimum square error and, at the same time, limit the dynamics of the weight vector. The weight vector is recalculated each time step, for more details see Benvenuto & Cherubini (2002); Guillemin et al. (1971).

Mean Weight

Because the weight vector generated by the LMS algorithm is very reactive to acceleration and deceleration processes, it is averaged by a running mean block that calculates the mean weight vector over the length of the situation. Both the mean and the LMS weight vector are combined into a slowed weight vector by multiplication with the parameter a or $1 - a$ respectively.

Weight Booster

The LMS algorithm adapts to new situations with a delay that is roughly its depth l plus the length of the prediction interval n, which equals the number of memory elements involved in the adaption. A change in velocity needs to pass through most of the memory elements before its effect becomes visible in the weight vector. For instance, for a car traveling in a city, the weight vector produces rather stable results while the car is traveling at a constant speed but it will react slowly to sharp braking or fast acceleration.

Table 1: Speed Prediction Parameters with example values used during development

parameter	example value	description
n	8	number of steps to predict
l	12	depth of FIR filters and dimension of weights vector
a	0.275	influence of the mean weights vector
b	2	influence of weight booster
c	0.2	boost limit
d	1	boost gain

The algorithm is designed to fix this problem by manipulating the weight vector in order to emphasize the most recent speed history elements to react more quickly to a spontaneous change in behavior: a length l base vector is multiplied with a scalar calculated from the slope of the velocity curve and is bounded above by the boost limit c. In order for the impact of the booster to remain present for a longer period, the generated "impulse" is broadened using a unity-weight FIR filter.

Parameters

The performance and precision of the speed predictor depends on some fundamental parameters that are summarized in Table 1. The given values are the result of some evaluations during the design phase based on few exemplary scenarios and should give a rough idea to start an implementation. However for a proper implementation a more thorough, numerical optimization is recommended but out of scope of this essay. It is important to note that all of the below parameters influence the prediction in a way that usually makes adoptions to all parameters necessary if one parameter is changed. In many cases more than one possibility exists that can lead to a desired result for one scenario, but looking at multiple scenarios usually only one if any of the possibilities lead to an overall improvement of performance.

Number of Steps to Predict (n)

This key parameter determines the number of steps to be predicted — n = 8 means the algorithm predicts the speed in 8 time steps. Obviously a higher value increases the prediction error, whereas lower values gives more precise predictions. The setting of this parameter is very important because its influence on the other parameters is tremendous, for instance a high value for n will on the one hand require a higher a and on the other hand require more influence of the weight booster, b. Also, this parameter is common with all components of the algorithm, so its influence has to be regarded globally. Different settings and their impact, especially on position prediction, are discussed in section 6.4.

Depth of FIR Filters (l)

The FIR filters' depth used in the speed predictor is a common value because all blocks share the weight vector. Also the parameter l is, unlike all other parameters mentioned here, a design time parameter that cannot be changed easily as it is hard-coded into the FIR filters and the constants. Nevertheless, its influence on the prediction should be discussed here. For the fact that the depth of a filter resembles its amount of memory elements, higher values for l give more stable and less reactive prediction results. Changes in the situation need more time to propagate through the memory elements, hence it takes a longer time to adapt to changes. Smaller values for l improve reaction time but also result in less stable and more fluctuating predictions that often overshoot at slight changes.

Influence of the Mean Weights Vector (a)

The weight vector in the standard case (disregarding the weight booster) is combined from the current weight vector produced by the LMS algorithm and its running average. Setting a to the maximal appropriate value a = 1 produces a very stable weight vector but also removes the direct influence of the LMS algorithm to the weight vector and thus the reactivity. This is caused by the fact that in this model, the mean weight vector is never reset and thus provides an "all time average".

Influence of the Weight Booster (b)

This parameter determines the overall influence of the weight booster. Higher values tend to produce overshoots as a trade-off to slow response to a change in situation if lower values are used. Generally all three values influencing the weight booster should be tuned according to the length of the prediction n. With high n, b should be increased because a faster reaction is necessary due to the latency of the LMS algorithm.

Boost Limit (c)

The "boost" vector, or more precisely its scalar values, are influenced by the slope of the velocity curve. The "boost limit" defines an upper bound to those scalar values.

Boost Gain (d)

This parameter multiplies the influence of the slope difference before the broadening and limiting of the pulse. Thus a higher value generates very quick increase once a steep slope is detected - in other words, it pushes the boost weight vector more quickly to the limit. Lower values produce a smoother response to steep velocity slopes.

Distance Calculation

The speed predictor predicts the vehicle speed some time steps ahead. The position predictor in turn requires as an input the distance to cover in the next time steps to calculate the future position. The most precise approach is to predict a velocity value for each time step in the prediction period and sum up the difference. Because this requires a set of n speed predictors which increases the computational efforts by n, a simpler approach is chosen in this implementation. Our algorithm uses the current speed and the predicted speed and calculates a linear approximation between the two. The distance to cover s is then the area under the speed curve for a duration of t (n time steps):

$$s = t \cdot \left[v_{current} + 0.5 \cdot \left(v_{pred} - v_{current} \right) \right]$$

Position Predictor

The position predictor uses the current position and direction of movement, digital map data as well as the predicted distance to cover as inputs and outputs a predicted position and its reliability. It is invoked once per time step and tries to first find the current road segment of the vehicle, then determines a number of possible prediction paths and finally chooses the most probable path and returns its end point.

Determine Current Road Segment

All known nearby road segments (taken from the digital map) are evaluated for the distance of the current position to the closest point on the respective road segment. Three criteria must be met in order for a road segment to be chosen:

- The distance to the closest point is smaller than a threshold.
- The absolute value of the difference between the direction of movement and the road segment's direction is not larger than $\pi/2$ because vehicles usually do not move perpendicular to streets.
- It is the closest road segment satisfying both criteria 1 and 2.

In case no road segment is found that fulfills all of the above criteria, the algorithm returns the current position as prediction result with an estimated accuracy of 0%. Possible causes range from wrong GPS positions during the initialization phase of the GPS device and inexact map material to driving or parking on streets or private property that is not (yet) included in the map material.

Determine Possible Paths

First, the remaining distance from the current position to the respective end point of the road segment sr is calculated. If the road segment's end point is further away than the distance to cover (sr $\geq$ s), the predicted position will be located between the current position and the road segment's end point. Therefore, the predicted position is determined along the road segment's polyline towards the end point, covering the given distance s. In the case that the remaining distance sr is smaller than the distance to cover (sr < s), the predicted position is moved to the road segment's end point and that distance is subtracted from the remaining distance. Subsequently, the next road segment of the prediction path is determined. If the mission of the vehicle is known in advance, the next road segment is chosen according to that mission. Otherwise, in order to find the next road segment, the number of possibilities is determined from the digital map: at a junction, all connected street segments are considered possible candidates. The current road segment, however, is not considered as an alternative — in other words, the vehicle is not expected to u-turn. Three cases exist:

- No candidates exist, so the current road segment ends in a node that has no other road segments referenced. In this case the relative probability of the current path is decreased in the relation to the amount of distance already covered.

- One candidate means there is no choice and the vehicle is moving along a road without an intersection at the current node. Determining the next road segment and updating the path accordingly is trivial.

- Multiple candidates are available, so the predicted path hits some kind of intersection. Hence the process of determining the next road segment becomes a bit more complex: Initially, all candidates must be assumed to be equally probable.

The procedure is repeated recursively until all distance s is covered and all possible paths of length s have been determined (effectively yielding a tree of possible road segments, with leaves at all possible future locations).

Pick Best Path

It may be desirable for an application that the prediction comprises all possible future realizations. However, if the prediction routine should return the future position along only one predicted path, the best of the alternative paths found must be chosen. If the mission of an observed vehicle is unknown to the algorithm, it must more or less issue a guess as to what option the driver of a car will go for. The range of alternatives is narrowed down in three steps:

- Estimated probability: In the current implementation of the algorithm, this first step will only remove the paths that end in a dead end and hence have a reduced relative probability. All other paths are considered equally probable and hence cannot be classified by their probability. For instance, the car hits an intersection with three alternatives, one of which being a dead end street. The dead end would be removed from the candidates, whilst the other two possibilities are equally probable.

- Way Changes: The number of street changes is the primary decision criterion for the algorithm. It is assumed that in case multiple paths exist, the driver stays on the current street. Hence the path with the least number of street changes is favored for the prediction. It is furthermore assumed that if it is necessary to change the road at some stage in all paths, the driver still stays on the current road as long as possible.

- Direction Difference: Should the way change criteria be unable to choose one candidate, the total difference of direction along the path is considered. Assuming the driver to be lazy, the path encountering the least change in direction is chosen to be the best path.

Clearly, criteria (b) and (c) do not increase the probability of a certain path. These are merely decision criteria in order to choose one path from multiple options. Choosing a random path statistically produces the same error, but has a severe disadvantage in terms of continuity: as the algorithm is executed each time step, it should return consistent values from one step to the next; when using a random selection, it is most likely that the algorithm will return a completely different position each time it is invoked. From a statistical point of view, this does not change much but for another program or algorithm that is based on the results of the prediction it may very well change things depending on the application. For the very same reason, it is very important for the algorithm to return the estimated probability of a given prediction because another program can then classify the prediction accordingly.

CHANNEL PREDICTION

It is clear from Equation 4 that the network's future connectivity does not only depend on the future vehicles' positions. An adequate channel model has to be chosen and constantly updated to reflect the changing radio environment. In

Equation 1, we have shown an exemplary simple channel function, the disc model. We shall call two nodes i and j connected if the path loss $\beta(d)$, a function of the distance between the two nodes, does not exceed a certain threshold β_t :

$$\beta_t \overset{!}{>} \beta(d) = 20 \log_{10} \frac{4\pi d_0}{\lambda} + 10\alpha \log_{10} \frac{d}{d_0}$$

(5)

The path loss consists of two components: a constant added that reflects the loss related to the wavelength of the signal and a distance-dependent term that represents the propagation of the radio wave through space and the resulting diminishment of power due to the growth of the wave sphere's surface. Due to various propagation effects, the path loss exponent (PLE) α is variable, to reflect various environments' radio properties (and usually ranges between 2 and 3). To account for reflections, scattering and shadowing, an additional stochastic variable β_s is introduced that is log-normally distributed with zero mean and variance σ^2:

$$\underbrace{\beta(d)}_{\text{measured}} = C + \underbrace{10\alpha}_{\text{unknown}} \log_{10} \underbrace{\frac{d}{d_0}}_{\text{known}} + \underbrace{\beta_s}_{\text{unknown}}$$

(6)

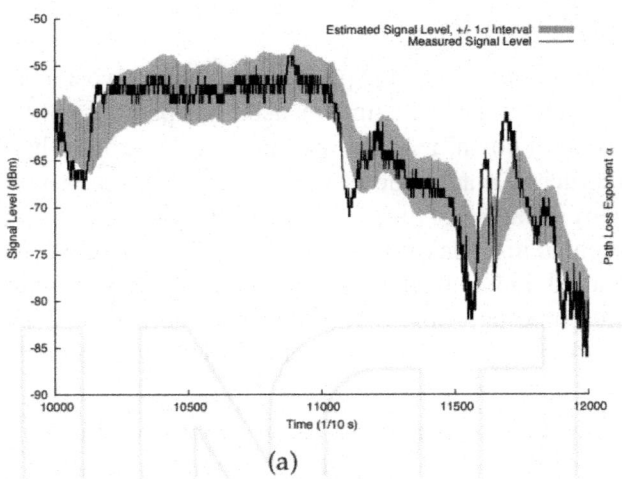

(a)

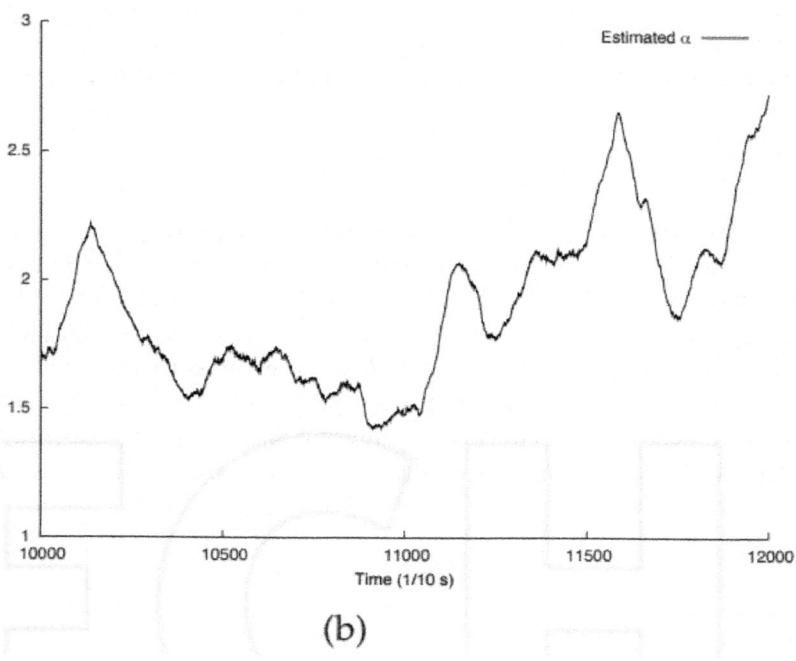

(b)

Figure 3: Parameter estimation using particle filter: measured and estimated signal level, estimated PLE.

Through constant exchange of position messages, two nodes can determine their distance d and at the same time measure the path loss $\beta(d)$ between them (terms marked as "known"). Assuming that βs is log-normally distributed and knowing the order of magnitude of the variance, we suggest to use a particle filter for online estimation of the PLE α and subsequent prediction, analogous to the work presented in Rodas & Cascon (2010). To study the vehicular channel, we have recorded and evaluated several hours of measurements. Figure 3(a) shows the measured path loss (solid line) and the path loss computed using the estimated PLE plus βs's 68% (one standard deviation) confidence interval (filled curve).

The standard deviation of the measured signal level was estimated around 3 dB, the system constant C was -42 dB and the duration of the displayed dataset is 20 seconds. The estimated PLE is shown in Figure 3(b). For connectivity prediction, we propose to employ the same concepts as used for position prediction to at least estimate the trend of the PLE. Furthermore, as we have discussed in the section on related work, it is very important to distinguish between LOS and NLOS conditions.

In Nagel & Eichler (2008), we have introduced a method for V2V channel simulation in environments that include objects that possibly obstruct a direct LOS path and have discussed how a dual-slope channel model can be implemented to account for these objects. Consequently, we propose to complement the path loss formula in Equation 6 accordingly and incorporate the information on buildings and other obstacles from the digital map material (that is already used in the position prediction) to determine if the path between the predicted future vehicle positions is LOS or not.

The propagation breakpoint derived from the map should then be accounted for in the PLE estimation. In simulations, we have realized quite accurate channel parameter estimations using a particle filter that accounts for LOS and NLOS conditions, based on information about surrounding radio obstacles. It is, however, clear that the effects of large-scale fading have a large impact on the future connectivity but are hard to account for. Due to positioning errors, it is virtually impossible to predict small-scale fading effects. When evaluating the connectivity matrix computed from the predicted positions and the predicted channel, additional information about the reliability of the prediction should be provided and accounted for.

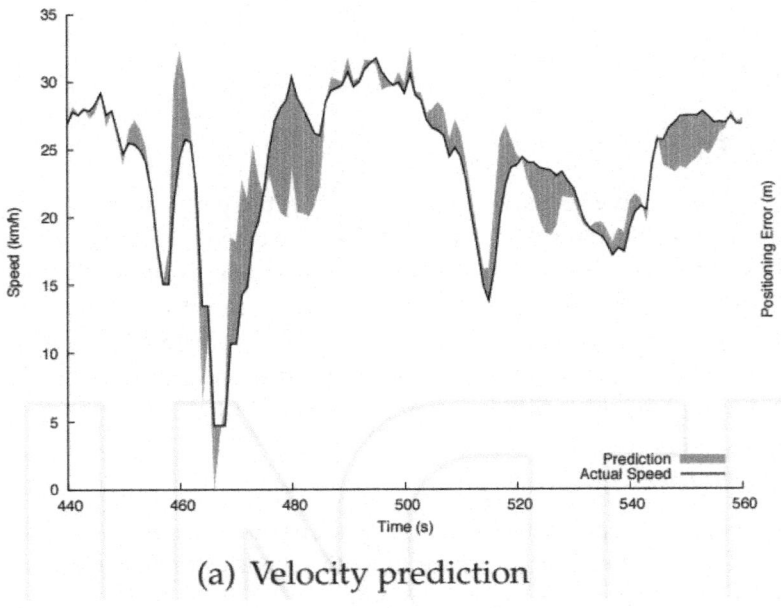

(a) Velocity prediction

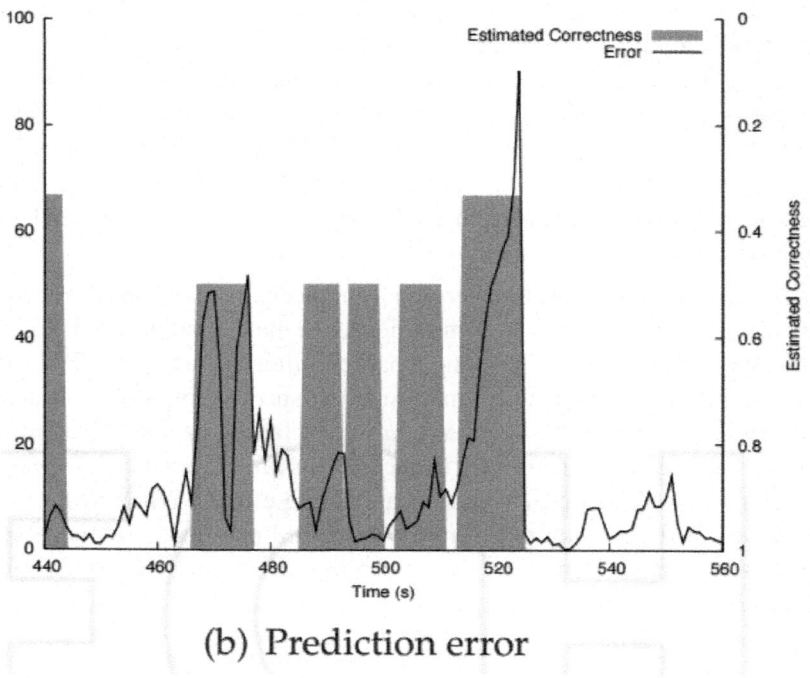

(b) Prediction error

Figure 4: City Scenario.

RESULTS

Three representative scenarios were chosen for the performance evaluation of the developed algorithm. These scenarios are based on GPS tracks downloaded from the OSM portal that were selected to provide maximum diversity in the results presented below: the chosen tracks were recorded in a city as well as in suburban and highway surroundings. We have evaluated the three scenarios concerning the accuracy of both the speed prediction and the resulting predicted position under the three different environmental settings.

City Scenario

The first data set represents a typical city scenario. It has been recorded in the german city of Herne in the Ruhr area, with speeds of up to 50 kilometers per hour and a total length of about 20 minutes.

Figure 4(a) shows a section of the actual vehicle speed (solid line), the area of the filled curve reflects the velocity prediction error where the upper or lower edge of the area marks the predicted speed. For easy comparison, the predicted values are shifted by 8 seconds, so that the real value and the value that has been predicted for that instant are matched in time. The results show that especially at steep slopes, the algorithm overshoots significantly and predicts too low or too high velocity values. This could be tuned using the parameters of the speed prediction in order to achieve better performance in the particular scenario, but the impact on other scenarios is hard to estimate and thus requires significant research efforts. Figure 8(a) shows the distribution function of the position prediction error in meters (upper blue line). The mean error is 13.4 meters, the median amounts to 7.5 meters. Figure 4(b) shows a section of the prediction error over time; the filled curve represents the estimated probability (correctness) of the prediction. Note that the second Y axis has been reversed for better readability.

As explained in section 4.5, the estimated probability generally equals 1 if the position predictor identifies only one possible path for the vehicle, based on the map data. In the presented case that the mission or route of the car are unknown to the algorithm, the choice of the path used for prediction is arbitrary once it encounters p multiple possible paths. Hence the estimated probability drops to 1/p. This, in turn, means that if a large error occurs while the estimated probability is 1, an error in the position predictor or the map material should be assumed, while high prediction errors with low estimated probability are most likely to be produced by the fact that the mission of the car is unclear. Should the position predictor be either unable to find the current segment or to find any path to continue, the estimated prediction probability drops to zero. This is the case at the beginning of the city scenario and is the result of pulling out of a private parking area in a sort of backyard. As there is no map material covering that area, the position predictor cannot give useful predictions based on the fact that it is unclear on which road the vehicle is driving. In such situations, the position predictor simply returns the current position.

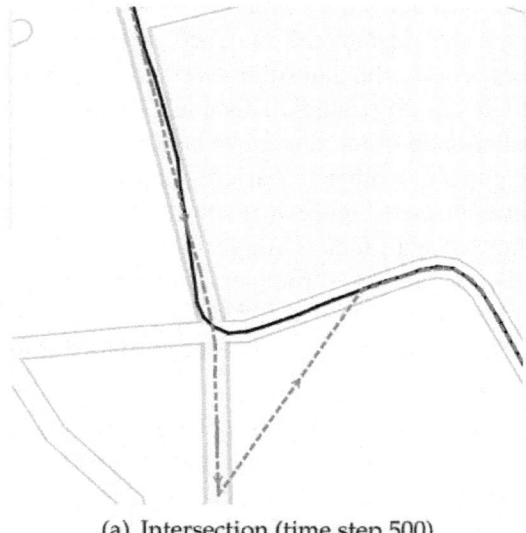

(a) Intersection (time step 500)

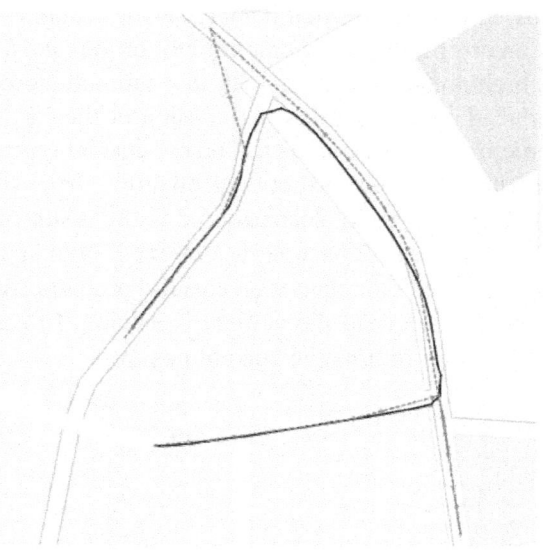

(b) T-Intersection (time step 200)

Figure 5: Scenario 1 - Prediction Behaviour at Crossroads.

To get an idea of the nature of an error, it is helpful to visualize the real and the predicted path as shown in Figure 5. The actual path is shown as a black solid line while the predicted path is shown as a dashed orange line.

Around time step 520, Figure 5(a) shows a typical situation in which the prediction algorithm encounters a crossroads. For the reasons explained in Section 4.5, the algorithm always prefers to choose the path that continues on the current road. This can be seen in the Figure where the predicted green dots continue straight on, while the real, blue trace turns onto the intersecting road. Once the car is on the new road, the prediction adapts to the new situation and continues its prediction along the new road. This can also be seen in Figure 4(b), where the error peaks due to the large discrepancy between the real and the predicted position. At the same time, the estimated probability drops to 0.33, due to the fact that the algorithm recognizes three alternative paths at the intersection. Figure 5(b) shows a similar example around time step 200 as the car moves towards a T-crossing. The algorithm chooses the path with the lowest total change in angle, which in this case is the wrong choice. This, again, results in a drop of the estimated probability and a peaking error.

As can be seen from the examples above, the high error in urban scenarios is widely based on the fact that usually many intersections lie along the path and high prediction errors are introduced if the algorithm chooses the wrong path. We have argued that this fact can be significantly improved if the cars mission is known to the position predictor and, consequently, the correct path can be used for the prediction.

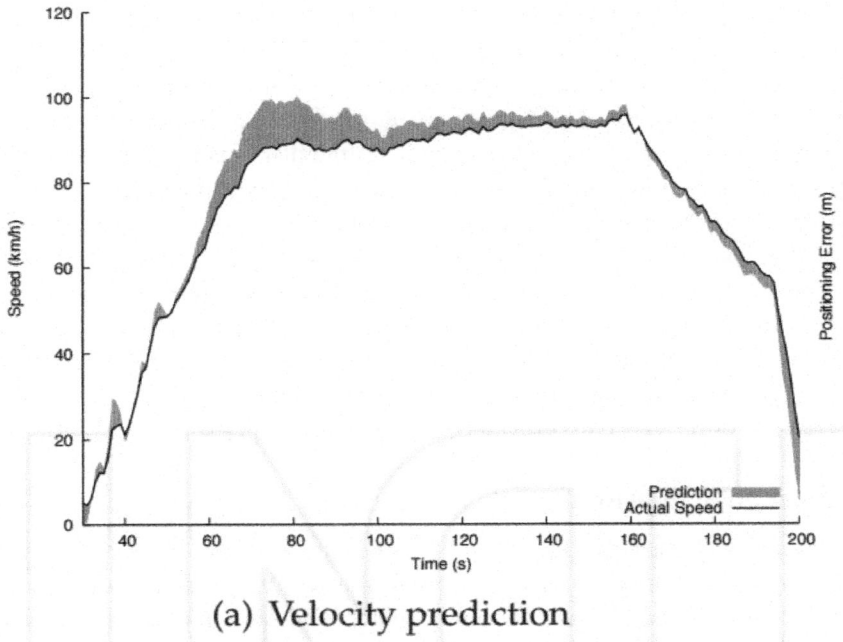

(a) Velocity prediction

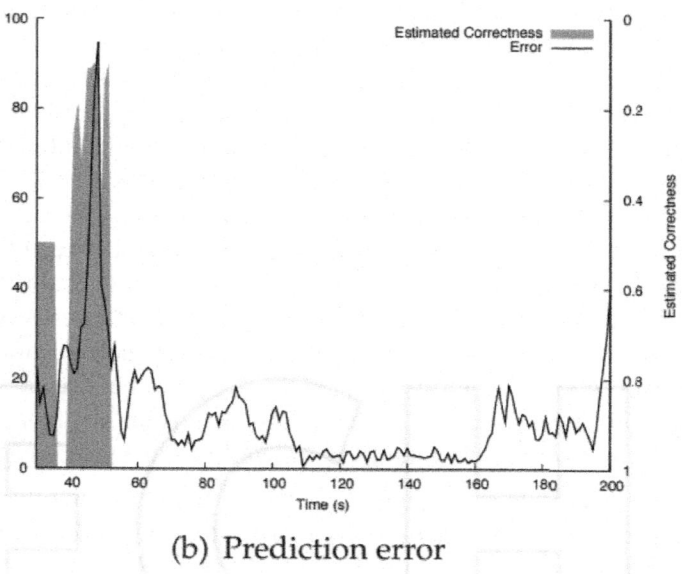

(b) Prediction error

Figure 6: Suburban Scenario.

This also implies that the mean prediction error is not a very good metric in order to assess the precision of the prediction. The algorithm may predict the cars position with an error of less that five meters in cases where there are no intersections, but at each intersection an error of up to 50 meters is possible that will greatly influence the mean error. A more suitable metric to identify the amount of such situations in the scenario is the discrepancy between the mean error and the weighted mean error that includes the estimated probability. The weighted mean error is simply the mean over the prediction error, multiplied with the estimated probability. Because the estimated probability drops at intersections, a prediction error in this situation is weighted less than an error occurring along a straight, intersection-free road. The weighted mean error in this situation amounts to about 7.4 meters, which is significantly less than the mean error of 13.4 meters and therefore shows that the prediction quality could be greatly improved with knowledge of the cars future route.

Suburban Scenario

The second scenario covers the suburban area of Wiener Neustadt in Austria. The driver first hits the B17 road and afterwards enters the suburban area where the car is parked at a shopping center. The section of the velocity graph in Figure 6(a) shows a short drive to the highway like state road. The speed prediction is rather stable whilst traveling at constant speeds between second

50 and second 200. This is accompanied by a rather small prediction error as shown in Figure 6(b). A peak in the prediction error at second 40 occurs due to a wrong choice of the next segment - again based on the fact that the cars mission is unknown. The fact that the algorithm had the choice of multiple directions is visualized by a significant drop of the estimated correctness curve. Due to a very sharp deceleration around second 200, Figure 6(b) shows another peak without a drop in probability for the fact that no other possible directions are identified. The reason for the braking is unclear, an explanation cannot be found in the map material nor the satellite image of the area. Figure 6(b) also shows a few more peaks based on decisions for the wrong directions and braking actions. The distribution function shown in Figure 8(b) (upper blue line) is slightly flatter than the one from scenario 1. The median error (about 10 meters) is slightly higher than in the city, probably caused by the fact that either map material or the GPS device used to record the track are less precise than in the city scenario. The 90% percentile is significantly smaller (22 meters compared to 34 meters), which supports the before assumption and leads to the conclusion that the overall position prediction is better in the suburban scenario. The mean error of 13.5 meters, however, is almost identical to the city scenario.

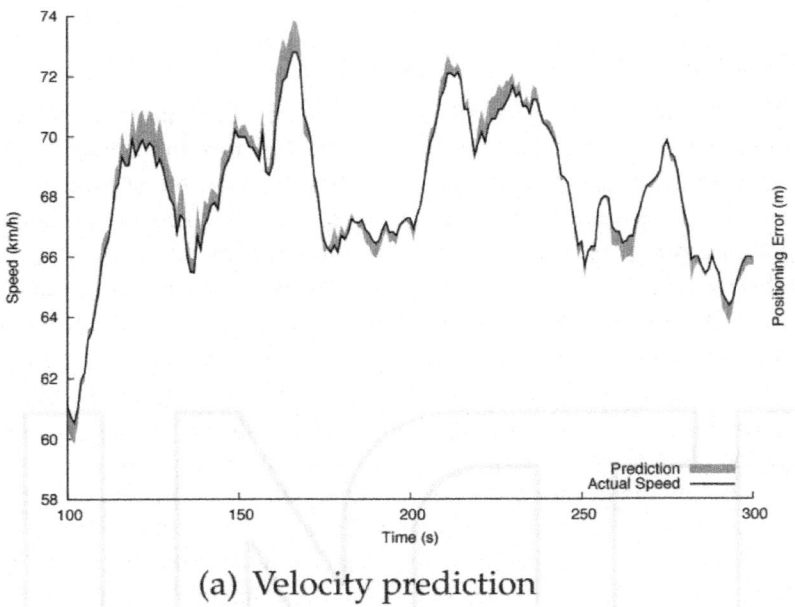

(a) Velocity prediction

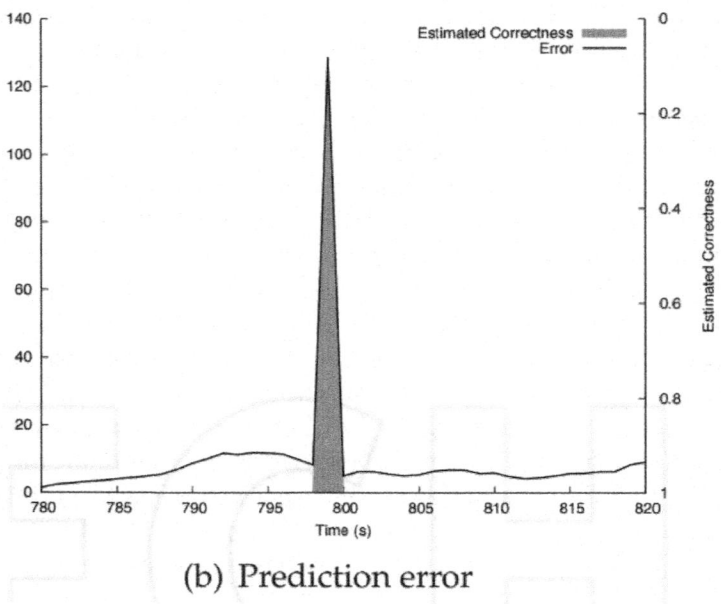

(b) Prediction error

Figure 7: Highway Scenario.

Highway Scenario

The third scenario covers a ride on a highway near the English town of Cambridge during rush hour traffic, which explains the unsteady velocity curve shown in Figure 7(a). Again, a representative section has been chosen for presentation. The prediction error plotted in Figure 7(b) shows constantly low prediction errors. From the distribution function shown in Figure 8(c) (upper blue line) reveals a very steep slope, which means excellent performance of the algorithm as more than 95% of all errors are below 10 meters. The mean error amounts to only 4.6 meters and the median is 3.8 meters. One notable peak of the error, accompanied by a drop in the estimated probability occurs around time step 800 (see Figure 7(b)). At the instant when the car is crossing a highway bridge the algorithm wrongly chooses the current road segment to be part of the road below the bridge that shares a node with the highway. Consequently, the algorithm chooses the wrong road segment to continue prediction. This error is caused by an unfortunate combination of a small measuring error of the GPS device and an imprecision of the map material, where the highway shares a node with the intersecting road (although they are on different levels).

Another error source stems from the fact that in the used map material, a road is represented by a polyline, and all vehicles are assumed to be positioned on this line. In reality, highways consists of a number of parallel lanes that have a certain lateral displacement. Although a suitable representation of road lanes has been suggested, the data available today does not yet represent different lanes. We expect, however, that the error would be decreased if this information could be accounted for in the prediction.

Influence of Prediction Length

The results that we have discussed above were obtained through prediction with period lengths of eight seconds. This sections evaluates the same scenarios with longer prediction lengths of 16 and 32 seconds. The resulting position error distribution functions are shown in Figure 8. The steepness of the distribution functions decrease as the prediction length increases, due to less prediction accuracy across a longer period of time, i.e., the tendency to produce greater errors. The maximum error also increases dramatically, because in case that a wrong path is chosen, the prediction continues along the wrong path for a much longer time before it is corrected as the car turns the other way. The mean error is also shifted towards higher values with increasing prediction intervals.

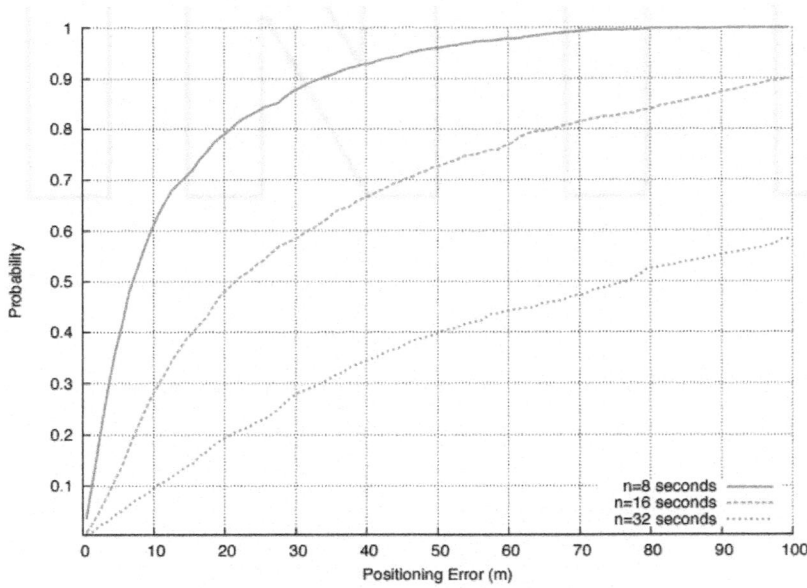

(a) Scenario 1 - City

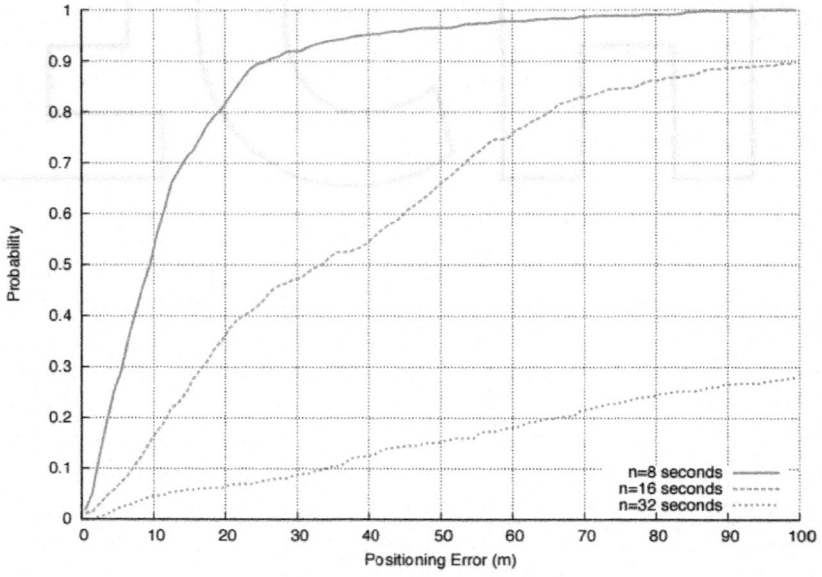

(b) Scenario 2 - Suburban

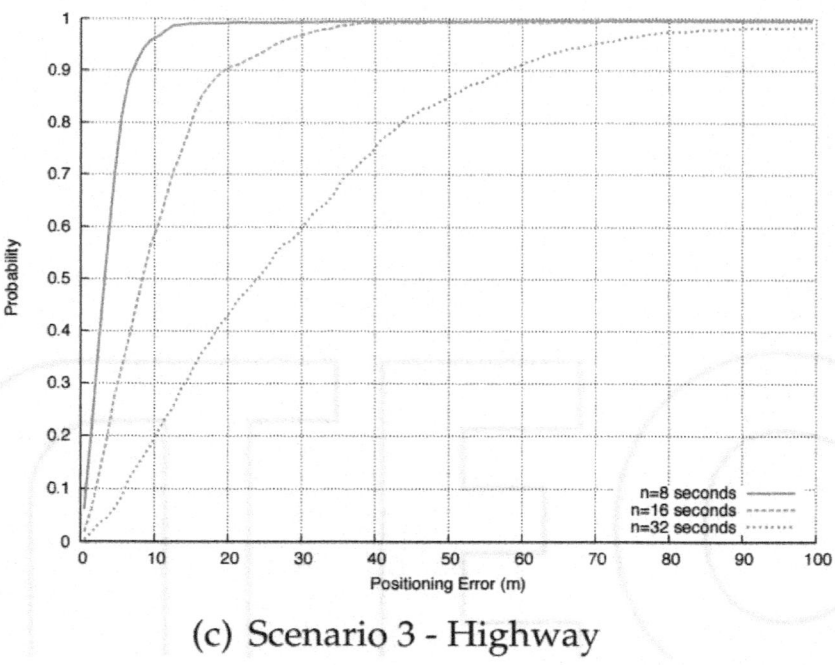

(c) Scenario 3 - Highway

Figure 8: Prediction Error Histogram - Different Prediction Lengths.

In the city and suburban scenarios, the prediction is already significantly less reliable with n = 16 time steps (i.e., seconds) as a prediction interval. It is rendered basically useless with a prediction interval of n = 32 and the majority of errors are out of scale of the histogram. The highway scenario, however, behaves much more stable as the prediction interval is increased and still returns useful results using a prediction horizon of 32 seconds. To a large extent, this is based on the relatively stable velocities and absent alternative paths along the way.

Path Loss Estimation Error

To determine the future connectivity of a network, a node has to predict the positions of all relevant vehicles (usually the one- or two-hop radio neighborhood) and determine the resulting path loss to decide whether it will be connected to this vehicle or not. The first error source of the path loss estimation is, of course, the error induced by inaccurate position prediction. Fortunately, this error term strongly depends on the distance d between the two involved vehicles. Let us assume that the estimating vehicle has perfect ego positioning and position prediction and let Δd denote the maximum positioning error. The resulting absolute estimation error range is then:

$$\Delta\beta(d, \Delta d) = \beta(d + \Delta d) - \beta(d - \Delta d)$$
$$= 10\alpha \log_{10}\left(\frac{d + \Delta d}{d - \Delta d}\right)$$

(7)

The second influence on path loss estimation is the accuracy of the predicted path loss exponent. Let $\Delta\alpha$ denote the PLE's maximum estimation error. The resulting absolute path loss estimation error is:

$$\Delta\beta(d, \Delta\alpha) = \beta(d, \alpha + \Delta\alpha) - \beta(d, \alpha - \Delta\alpha)$$
$$= 20\Delta\alpha \log_{10}\left(\frac{d}{d_0}\right)$$

(8)

Clearly, there exists a strong negative correlation between the path loss estimation error and the distance to the tracked vehicle, d. With increasing d, the implications of prediction errors become less important with respect to the path loss. The situation is contrary regarding the PLE estimation: because the relation is linear, a large path loss estimation error results if the distance d to the tracked vehicle increases. The problem is that Δd and $\Delta\alpha$ are not known at runtime; therefore, we suggest to keep the prediction results and constantly compare them against the predictions to obtain a statistic of the errors. This information should consequently be used to determine a prediction's reliability.

CONCLUSIONS AND OUTLOOK

In this chapter, we have presented an algorithm for the self-adaptive prediction of mobile nodes' future positions. The algorithm is targeted at vehicular applications with nodes that move along the road grid, of which a digital map is available at runtime. We have introduced the necessary building blocks along with their parameterization, discussed some performance studies and pointed out individual strengths and shortcomings. Three exemplary scenarios have been studied: city, suburban, and highway. Given a prediction interval of eight seconds, the algorithm performed well in all of the scenarios, resulting in a mean prediction error of only about 14 meters. On a highway, the mean error is less than 5 meters. As the prediction interval increases, the performance of the algorithm degrades significantly in the city and suburban scenario. On a highway, however, the mean error is around 30 meters for an interval of 32 seconds which may still be acceptable, depending on the application. We have already argued in the discussion that the position prediction accuracy in the city and suburban scenario is mainly degraded due to an incorrect path selection as the considered vehicle approaches an intersection. In our studies, the future path has been selected randomly from the set of possible paths.

The necessary assumptions have been explained in Section 4.5.2. If the information is available, we strongly propose to consider vehicles' missions for path prediction. Considering the city scenario, if only those prediction errors are evaluated for which the path selection is correct (i.e., the predictor estimates the correctness as 1), the mean error can be decreased to about 8 meters, the median to about 5 meters. Taking the mission into account, these extremely low errors seem feasible. The computation of the distance to cover is currently calculated using the area under a linear graph between the current velocity and the predicted velocity, thus assuming a linear acceleration. Some thoughts should be given to a substitution of this simple approximation with a more sophisticated implementation. One idea that is rather complex in terms of computational efforts is to use n velocity predictors to predict a velocity for each time instant and thus removing the interpolation. Another approach could be to model the acceleration and deceleration behavior of a typical driver. If the vehicle is autonomous, reproducing the design of the longitudinal controller could increase the prediction performance. Velocity prediction, too, offers some optimizations opportunities. The key measure necessary here is a numerical optimization of the parameters a, b and c mentioned in Table 1 over a large number of scenarios of adequate length. Appropriate parameter sets could be computed beforehand (and even optimized online) and information about the current driving situation could be used to select the most suitable set.

This selection, in turn, could be used to provide other applications with valuable information about the current environment. It is expected that an adoption of these parameters will lead to a somewhat significant improvement of the speed prediction. An urban scenario requires much quicker reactions to speed changes and thus needs more contributions and stronger influence of the weight booster than a highway scenario. The highway scenario, in turn, profits from a more stable prediction based to a large extent on the mean weight vector and requires virtually no influence of the weight booster. Longer-term prediction of the wireless channel still imposes the largest problem when it comes to evaluate the future connectivity from predicted positions. Further work is necessary to evaluate the dynamics of the radio channel and design an appropriate predictor. We propose to include further information about the environment (see above) in order to distinguish between different surroundings and consequently adjust the appropriate channel parameters (such as the variance of the large-scale fading). An interesting idea in this context is to share and aggregate knowledge of the communication channel obtained from measurements between nearby vehicles. Another situation that requires attention is the channel prediction for vehicles that are actually outside of a node's communication range and channel parameter estimation is obviously not possible. In this case, the channel has to be estimated from measurements conducted with and from nearby vehicles.

REFERENCES

1. Althoff, M., Stursberg, O. & Buss, M. (2010). Computing reachable sets of hybrid systems using a combination of zonotopes and polytopes, Nonlinear Analysis: Hybrid Systems 4(2): 233 – 249.

2. Benvenuto, N. & Cherubini, G. (2002). Algorithms for Communications Systems and Their Applications, Wiley, New York.

3. Boukerche, A., Rezende, C. & Pazzi, R. (2009). Improving neighbor localization in vehicular ad hoc networks to avoid overhead from periodic messages, pp. 1 –6.

4. Cheng, L., Bai, F. & Stancil, D. (2009). A new geometrical channel model for vehicle-to-vehicle communications, pp. 1 –4.

5. Guillemin, E. A., Kalman, R. E., DeClaris, N. & Andersen, J. (1971). Aspects of network and system theory, Holt Rinehart and Winston, New York.

6. Huang, C.-J., Chuang, Y.-T., Yang, D.-X., Chen, I.-F., Chen, Y.-J. & Hu, K.-W. (2008). A mobility-aware link enhancement mechanism for vehicular ad hoc networks, EURASIP J. Wirel. Commun. Netw. 2008: 1–10.

7. Kaaniche, H. & Kamoun, F. (2010). Mobility prediction in wireless ad hoc networks using neural networks, Journal of Telecommunications 2(1).

8. Matolak, D. W., Sen, I. & Xiong, W. (2006). Channel modeling for v2v communications, Mobile and Ubiquitous Systems - Workshops, 2006. 3rd Annual International Conference on, pp. 1 –7.

9. Maurer, J., Fugen, T., Schafer, T. & Wiesbeck, W. (2004). A new inter-vehicle communications (ivc) channel model, Vol. 1, pp. 9 – 13 Vol. 1.

10. Molisch, A., Tufvesson, F., Karedal, J. & Mecklenbräuker, C. (2009). A survey on vehicle-to-vehicle propagation channels, Wireless Communications, IEEE 16(6): 12 –22.

11. Nagel, R. (2010a). Altruistic traffic limits computation in wireless broadcast networks, Proceedings of the Third Internation Conference on Advances in Mesh Networks.

12. Nagel, R. (2010b). The effect of vehicular distance distributions and mobility on vanet communications, Proceedings of the IEEE Intelligent Vehicles Symposium.

13. Nagel, R. & Eichler, S. (2008). Efficient and realistic mobility and channel modeling for vanet scenarios using omnet++ and inet-framework, Simutools '08: Proc. of the 1st international conference on Simulation tools and techniques for communications, networks and systems & workshops, ICST, pp. 1–8.

14. Paier, A., Karedal, J., Czink, N., Dumard, C., Zemen, T., Tufvesson, F., Molisch, A. F. & Mecklenbräuker, C. F. (2009). Characterization of vehicle-to-vehicle radio channels from measurements at 5.2 ghz, Wirel. Pers. Commun. 50(1): 19–32.

15. Rezende, C. G., Pazzi, R. W. & Boukerche, A. (2009). An efficient neighborhood prediction protocol to estimate link availability in vanets, MobiWAC '09: Proceedings of the 7th ACM international symposium on Mobility management and wireless access, ACM, New York, NY, USA, pp. 83–90.

16. Rodas, J. & Cascon, C. J. E. (2010). Dynamic path-loss estimation using a particle filter, International Journal of Computer Science Issues 7(3).

17. Samaan, N. & Karmouch, A. (2005). A mobility prediction architecture based on contextual knowledge and spatial conceptual maps, Mobile Computing, IEEE Transactions on 4(6): 537 – 551.

Chapter 10

MEASURING SCIENTIFIC MISCONDUCT—LESSONS FROM CRIMINOLOGY

Felicitas Hesselmann [1], Verena Wienefoet [1], and Martin Reinhart [1,2]

[1]Institute for Research Information and Quality Assurance, Schützenstraße 6a, 10117 Berlin, Germany

[2]Humboldt-Universität zu Berlin, Institut für Sozialwissenschaften, Unter den Linden 6, 10099 Berlin, Germany

ABSTRACT

This article draws on research traditions and insights from Criminology to elaborate on the problems associated with current practices of measuring scientific misconduct. Analyses of the number of retracted articles are shown to suffer from the fact that the distinct processes of misconduct, detection, punishment, and publication of a retraction notice, all contribute to the number of retractions and, hence, will result in biased estimates. Self-report measures, as well as analyses of retractions, are additionally affected by the absence of a consistent definition of misconduct. This problem of definition is addressed further as stemming from a lack of generally valid definitions both on the level of measuring misconduct and on the level of scientific practice itself. Because science is an innovative and ever-changing endeavor, the meaning of misbehavior is permanently shifting and frequently readdressed and renegotiated within the scientific community. Quantitative approaches (*i.e.*, statistics) alone, thus, are hardly able to accurately portray this dynamic phenomenon. It is argued that more research on the different processes and definitions associated with misconduct and its detection and sanctions is needed. The existing quantitative approaches need to be supported by qualitative research better suited to address and uncover processes of negotiation and definition.

INTRODUCTION

Studying a problem scientifically involves measuring things quantitatively [1]. Approaching a problem quantitatively promises to establish two foundations from which further research will be possible and justifiable. First a valid and reliable measurement shows the extent of the problem and helps estimating whether more scientific attention is warranted. Second, knowing the extent of a problem from scientific measurement is helpful in convincing other scientists and the public that more research on the problem is needed and should be supported (intellectually and financially). This strategy is as frequently used to introduce research on scientific misconduct as on many other topics. However, our argument will be that especially for research on scientific misconduct this strategy is problematic. It is problematic because statistics on misconduct are prone to biases well-known from Criminology, which find little attention in the literature on scientific misconduct. The measurement of the prevalence of retracted papers will serve as a case in point. The aim of this paper is, thus, to present an argument why measuring scientific misconduct quantitatively should not be first on our research agenda. Decades of research in Criminology and the Sociology of Deviance will serve as a background to support this argument.

MEASURING MISCONDUCT

Researchers seeking to appraise the prevalence of scientific misconduct try to do so by either looking at known instances of fraud and error, such as retracted articles, or by conducting surveys of scientists asking about their experience with misconduct. These strategies are comparable to the methods and resulting statistics in Criminology that are used to uncover the unknown "dark figure" of crimes in a population. Both the rate of retractions and the number of scientists admitting to or suspecting their colleagues of misconduct are subject to systematic errors. Because Criminology, with its long-standing tradition of investigating these errors [2], has produced notable insights into the problems associated with these methods, the knowledge gained by criminological research will serve as a reference point for our further argument.

Calculating some form of a retraction rate to assess the prevalence of misconduct constitutes a prevailing line of research [3,4,5,6,7,8,9,10,11,12,13]. Many studies focus on data of retracted articles retrieved via databases such as the Web of Science, PubMed, and the like. As with a crime rate, these observed data do not accurately characterize the true rate of scientific misconduct [2,14], but are the joint outcome of a number of distinct data generating processes that each produce a characteristic distortion (see also [15,16]). For the retraction rates, there are at least four different processes that need to be separated:

First, there is the actual occurrence of misconduct, the process many studies seek to explain or measure and that will result in fraudulent publications. Drawing an analogy to Criminology, this process represents the occurrence of criminal acts, which is at the heart of many theories trying to explain why some people resort to crime while others do not [2].

Second, there is the process of detection of misconduct. This process entails arising suspicions, allegations, and subsequent investigations and is comparable to the reporting of crimes to the police and the subsequent recording of these crimes by the police [15]. Thus, in addition to the crimes that form the official police statistics, there are infractions that are never detected by anyone, crimes detected but not reported, and crimes reported but not recorded by the police [14,15,17]. Hence, crimes officially recorded by the police pose only a fraction of all crimes committed. Moreover, the crimes detected and recorded might misrepresent the structure of incidents, because some crimes or forms of misconduct can be more difficult to detect, minor infringements might simply be shrugged off and show up nowhere in the data [14,15,18], or because some people are more successful at obscuring their deeds or intimidating possible witnesses. In science, there is evidence that many researchers who witness scientific misconduct refrain from reporting the incident [19,20,21]. To our knowledge, no research has investigated whether reported infractions differ systematically from unreported infractions.

The third process is the retraction of a publication, which can be compared to the sentencing in criminal procedure. There are many factors that determine the outcome of a trial, of which the actual crime is only one. For the case of scientific misconduct, many factors associated with the decision to retract an article (or not) remain unclear: A number of studies show that many journals do not have clear policies for dealing with misconduct [22,23,24,25] and handle allegations inconsistently and mostly on a case-by-case basis [26,27,28]. Additionally, even though resources such as the Committee on Publication Ethics (COPE) guidelines do exist, a notable number of editors are unaware of these guidelines [27]. Even in cases when a finding of misconduct has been established by the Office of Research Integrity (ORI), some affected articles are neither retracted nor corrected [29,30,31,32]. Criminological research shows that the aggregate rate of incarceration in a country varies independently from the rate of crimes recorded by the police [33] and exhibits varying and mostly inconsistent patterns of correlation with other measures of crime [34]. It seems optimistic to assume that this would not hold for scientific misconduct and to conclude that the development of retractions was directly relational to the development of misconduct. Many researchers thus acknowledge the fact that there is no way of knowing whether for instance, retractions have risen over the past decades due to a rising prevalence of misconduct or a rising awareness and scrutiny.

The fourth process that ultimately influences the records of retractions and is often overlooked is the way in which journals and databases identify retracted publications once an article has been retracted. To a lesser extent, data entry errors or misidentification also influence official police records [17], but rigid regulations and guidelines for record-keeping and considerable efforts for harmonizing international data (e.g., by EUROSTAT) make this problem less severe for criminological research as compared to research on scientific misconduct. There can be major inconsistencies as to how retracted articles are marked in the databases and whether they are marked at all. It is currently far from clear how many of the publications that were formally retracted will show up as "retracted articles" in one of the databases and how many remain to appear unmarked and "valid". Moreover, some journals may label a retraction by adding the term "retraction" to the title, while others may use different terms [29]. The number of retracted articles will, hence, vary greatly by the search term researchers use to locate retractions in the database.

We can, thus, learn from Criminology that measuring rates of misconduct should separate at least four different processes related to misconduct. What we can measure depends on the actual occurrence of misconduct (1); the detection of misconduct (2); the sentencing of misconduct (3); and the recording of misconduct (4). The problem posed by the entanglement of these four processes reaches much further than just an underestimation of the prevalence of misconduct. As every process systematically influences the probability that a given instance of misconduct will be selected into the sample of studied retracted articles, this selection bias will result in biased estimates in a regression analysis [35]. Because these selection rules are currently unknown, both the size, as well as the direction, of this bias remain unknown, rendering the analyses mostly meaningless.

The concurrence of these processes alone makes inferences about scientific misconduct extremely difficult, however, the current measurements also suffer from the fact that misconduct seems to be a moving target. In the absence of a generally valid definition of misconduct, both official institutions [36] and researchers apply a variety of definitions with differing scopes. Hence studies of scientific misconduct often do not measure the same thing. Looking at the analysis of retractions, researchers differ, e.g., as to whether they include plagiarism as a form of misconduct [7,11,37] or not, rather defining it as a form of error [12,38]. This inevitably results in differing numbers of retractions due to misconduct (see also [39]). Accordingly, numbers of retractions are not comparable across studies. Differing definitions of prohibited acts also trouble criminological research; hence many researchers consider it futile to compare raw crime rates across countries or even within a country over long periods of time. Growing (or declining) rates of retractions may not only be caused by greater (lesser) scrutiny or propensity to act on the part of reviewers and editors [6], but also by a changing range of behaviors that are considered to be misconduct. Because there is so little information on what behaviors exactly prompt journals to retract articles [5,28], we do not know if and how this definition has changed over time.

In the current situation, given only the retractions as they appear in the databases it is impossible to untangle these influences to make any statement about the prevalence or structure of either misconduct, the risk of its detection, or the probability that detected misconduct is sanctioned. It is crucial to gain a better insight into these distinct mechanisms, both in the form of theoretical consideration and empirical evidence, before analyses based on the rate of retractions can be meaningful.

As one reaction to the outlined shortcomings of official measures of crimes, criminological research has turned to self-report measures to gauge the true extent of criminal behavior [40,41]. This strategy is also applied when examining scientific misconduct [42,43]. The problems associated with self-report measures of delinquency continue to be of great interest in Criminology, as they may result both in under-reporting [17,40,41] and over-reporting of misbehavior [17,40]. Respondents might not want to tell the truth about their actual involvement in criminal behavior [40], or they might simply have forgotten about some acts [17]. Accurate recall of own behaviors poses a very serious problem in the research of misconduct using self-reports: Questions pertaining to only vaguely defined periods (*i.e.*, "ever" [44]) or to very large time-spans (*i.e.*, "during the last 10 years" [45]; "since entering medical school" [46]) will generally not produce very reliable or, for that matter, comparable answers ([41]; see [47] for a short overview). The difference between self-reports and official records furthermore varies, among others, by number of arrests [40,41], gender, and type of offense [41], hence producing systematically biased estimates. Additionally, for the case of self-reports, the problem of definition becomes crucial. The phenomenon of over-reporting arrests is commonly seen as resulting from respondents misleadingly counting all contacts with the police as arrests, even if they were not officially arrested [17,40]. Particularly, with regard to scientific misconduct, the absence of a stable definition might result in different people reporting the same act under different names. What might seem as falsification of data in one discipline, might be considered only sloppiness or even accepted practice in another discipline. This problem becomes obvious in the study by Titus, Wells, and

Rhoades [20], where 24% of reported incidents did not actually meet the federal criteria of misconduct.

DEFINING MISCONDUCT

The problems associated with defining scientific misconduct thus need further elaboration. Following Becker that "deviant behavior is behavior that people so label" ([48]: p. 9) and that the simplest conception of deviant behavior is a statistical one which defines everything as different, which lies too far from the average [48], presents us with problems regarding definition on two different levels. One level is represented by observers of scientific misconduct (e.g., Science Studies or administrators) and the other by the daily practices of scientists. On both levels, labeling occurs with specific restrictions but not independent of one another.

For the level of observation of misconduct the most severe restriction seems to lie in the fact that the labels for misconduct must come from scientific practice itself. Politicians, administrators or Science Studies scholars lack authority to label a specific scientific practice as deviant since only scientists themselves possess the necessary qualifications to do so. We are not talking about forms of misconduct that can be defined rather easily like outright plagiarism or fabrication here but about the many routine scientific practices that operate in a "grey zone". For such practices to become problematic and labeled as deviant the scientific community on the level of scientific practice has to act first and only then are we able to use the same label on the level of observation. To give an example, before certain forms of image manipulation could be labeled as misconduct, e.g., by journals, funding organizations or the ORI the scientists themselves had, first, to engage in these practices and then, second, to negotiate what the community rejects as too much "beautification" [49]. On the level of observation, definitions of deviance are thus always dependent on scientific practices and, in some sense, too late [50]. This aspect raises problems especially for research on misconduct because it cannot rely on an external source for distinguishing misconduct from regular behavior. Furthermore, labels for misconduct vary from one discipline to the next, presenting Science Studies with a difficult task in establishing a precise and generalizable terminology [49,51]. This may, in part, explain the variance in terminology in research on scientific misconduct presented above.

When switching to the level of scientific practice and the ways scientists negotiate the permissible and non-permissible forms of behavior on a day-to-day basis, the picture becomes more complicated. Here, these problems of definition can pose restrictions for scientific practice in that the capacity to innovate may be affected. In Criminology a classical idea is that innovation often comes along with the rejection of institutional norms and the establishment of new forms of behavior which, over time and through negotiation, become permissible forms of behavior ([52]: p. 231). This is not to say that for innovation to occur norms must always be broken. However, there is a tradeoff between rigid control of behavior according to institutional norms and the possibilities for innovation. Especially for the case of science, where we expect scientists to deliver new and innovative knowledge, that we sometimes call "revolutionary", a context that does not allow for non-conformist approaches may stifle innovation. Only the resulting processes of negotiation separating the permissible from the non-permissible will establish what can count as new and regular facts, theories, and methods [53]. Again, this is not to say that plagiarism, fabrication, and falsification should be tolerated but rather that efforts to prevent or sanction misconduct may have the non-intended consequence of rendering certain forms of scientific innovation more unlikely.

One can, thus, see that the problems associated with defining scientific misconduct lead to questions of labeling and ultimately to the processes of negotiation in different scientific fields about the boundaries of permissible practice. The dynamics of these processes and their specificity to certain research fields should deter from prematurely committing to overly precise or general definitions of scientific misconduct. Otherwise, research on scientific misconduct may suggest a more objective picture of scientific practice than it actually can deliver. One conclusion that can be drawn from this and from the experiences from Criminology is that more qualitative approaches are needed to analyze the processes of occurrence, detection, punishment, and recording of scientific misconduct in the places where these processes happen.

The line drawn between scientific misconduct and accepted research practice can be seen as a divide not between two types of research but essentially between science and non-science. "Boundary work" [54] can be a useful concept to study the ways scientists separate the permissible from the non-permissible. Cases of (alleged) misconduct, like the one by Cyril Burt, a famous psychologist who was posthumously accused of having falsified and fabricated data, give a rich picture of the complex social processes that lead to questionable practices and the ways these are detected and labeled as misconduct. In the case of Burt, his studies on the cognitive capabilities of twins, arguing that intelligence is mainly determined by genes, seem to have highly unlikely samples and results. However, even within Psychology, consensus has not been forthcoming on whether this qualifies as misconduct or not. It seems that cases of possible misconduct can be highly contested because they stand for more than just the one case. They are always also cases where the limits of the cognitive authority of a discipline or science in general are questioned [55]. We would thus suggest to continue existing research from Science and Technology Studies dealing with boundary-work [54], which not only broaches the issue of scientific misconduct but also the demarcation between science and non-science ([56]: p. 16). This line of research, as well as the literature on retractions and scientific misconduct, could benefit from including criminological approaches.

CONCLUSIONS

Making use of an analogy to criminological research and criminal statistics, this article addressed the problem of confounding different processes that influence the final number of retractions that appear in scientific databases. Moreover it discussed the problem of changing and fuzzy definitions of misconduct that can hardly be represented by quantitative data.

Because currently very little is known both about the outlined processes of occurrence, detection, punishment, and publication of misconduct and about the negotiation of definitions within the scientific fields, raw numbers of retractions or self-reported misbehaviors are not very telling. Every process systematically influences the probability that a given instance of misconduct will be selected into the sample of studied retracted articles and causes amongst others biased estimates. They will most likely paint an inaccurate picture both of the amount and the characteristics of misbehaviors and the scientists committing them.

More research is needed to illuminate these intermediary processes. Because of the explorative nature of such research and because the objectives of research are dynamic and deeply embedded within structures of everyday action, reasoning, tacit knowledge and sense-making, the methods of choice at this point would have to be qualitative (see, e.g., [57,58,59]). The definitions of misconduct, for instance, cannot be addressed by imposing pre-established definitions on the phenomenon in question, as is necessary in statistical analyses, but call for an open-ended approach that aims at discovering rather than confirming the various meanings of the label "misconduct". Qualitative research can deliver an empirically grounded overview that should serve as a starting point for further quantitative analyses.

Moreover we argue to altogether shift the research focus to address the processes of negotiation at the various levels—in the scientific community (micro), in institutions like universities and journals (meso), and in more general public discourses (macro)—more prominently. Instead of being solely conceptualized as error or noise that will taint quantitative analyses and that more research simply seeks to partial out, it could be noted that, just like the processes influencing criminal statistics, they are "an aspect of social organization and cannot, sociologically, be wrong." ([14]: p. 734). Social processes that shape definitions of scientific misconduct are not just sources of error or noise for researchers who are trying to measure misconduct but should be seen as an important resource and topic for further research. In fact, it is these very processes of negotiating definitions and boundaries of scientifically correct and scientifically fraudulent (or erroneous) behavior that lie at the heart of the scientific enterprise and therefore should be a major focus of the study of scientific misconduct [54,60]. An analysis of these processes should not simply aim at producing more reliable statistics in the end, but seek to shed light on the structures and mechanisms of scientific knowledge production.

ACKNOWLEDGMENTS

We would like to thank Natascha Trutzenberg for her help with retrieving and managing the literature for this paper and Marion Schmidt for bibliometric advice. This paper originates from project No. 01PY13009 funded by the German Ministry of Education and Research. The content of this paper is solely the responsibility of the authors.

REFERENCES

1. Porter, T.M. *Trust in Numbers*; Princeton University Press: Princeton, NJ, USA, 1996.

2. Biderman, A.D.; Reiss, A.J. On Exploring the "Dark Figure" of Crime. *Ann. Am. Acad. Pol. Soc. Sci.* 1967, *648*, 1–15.

3. Budd, J.M.; Sievert, M.; Schultz, T.R. Phenomena of retraction. Reasons for retraction and citations to the publications.*JAMA* 1998, *280*, 296.

4. Davis, P.M. The persistence of error: A study of retracted articles on the Internet and in personal libraries. *J. Med. Libr. Assoc. (JMLA)* 2012, *100*, 184–189.

5. Decullier, E.; Huot, L.; Samson, G.; Maisonneuve, H. Visibility of retractions: A cross-sectional one-year study. *BMC Res. Notes* 2013, *6*.

6. Fanelli, D. Why growing retractions are (mostly) a good sign. *PLoS Med.* 2013, *10*, e1001563.

7. Fang, F.C.; Steen, R.G.; Casadevall, A. Misconduct accounts for the majority of retracted scientific publications. *Proc. Natl. Acad. Sci. USA* 2012, *109*, 17028–17033.

8. Foo, J.Y. A Retrospective Analysis of the trend of retracted publications in the field of biomedical and life sciences. *Sci. Eng. Ethics* 2011, *17*, 459–468.

9.   Furman, J.L.; Jensen, K.; Murray, F. Governing knowledge in the scientific community: Exploring the role of retractions in biomedicine. *Res. Policy* 2012, *41*, 276–290.

10. Grieneisen, M.L.; Zhang, M.; von Elm, E. A Comprehensive Survey of Retracted Articles from the Scholarly Literature.*PLoS One* 2012, *7*, e44118.

11. Nath, S.B.; Marcus, S.C.; Druss, B.G. Retractions in the research literature: Misconduct or mistakes? *Med. J. Aust.* 2006,*185*, 152–154.

12. Redman, B.K.; Yarandi, H.N.; Merz, J.F. Empirical developments in retraction. *J. Med. Ethics.* 2008, *34*, 807–809.

13. Steen, R.G. Retractions in the scientific literature: Is the incidence of research fraud increasing? *J. Med. Ethics* 2011, *37*, 249–253.

14. Black, D.J. Production of Crime Rates. *Am. Sociol. Rev.* 1970, *35*, 733–748.

15. Boivin, R.; Cordeau, G. Measuring the impact of police discretion on official crime statistics: A research note. *Police Q.*2011, *14*, 186–203.

16. O'Brien, R.M. Police productivity and crime rates: 1973–1992. *Criminology* 1996, *34*, 183–207.

17. Kirk, D.S. Examining the divergence across self-report and official data sources on inferences about the adolescent life-course of crime. *J. Quant. Criminol.* 2006, *22*, 107–129.

18. Agnew, R.; Brezina, T. *Juvenile Delinquency: Causes and Control*, 4th ed.; Oxford University Press: New York, NY, USA, 2012.

19. Fanelli, D.; Tregenza, T. How Many Scientists Fabricate and Falsify Research? A Systematic Review and Meta-Analysis of Survey Data. *PLoS One* 2009, *4*, e5738.

20. Titus, S.L.; Wells, J.A.; Rhoades, L.J. Repairing Research Integrity. *Nature* 2008, *453*, 980–982.

21. Pryor, E.R.; Habermann, B.; Broome, M.E. Scientific misconduct from the perspective of research coordinators: A national survey. *J. Med. Ethics* 2007, *33*, 365–369.

22. Atlas, M.C. Retraction policies of high-impact biomedical journals. *J. Med. Libr. Assoc. (JMLA)* 2004, *92*, 242–250.

23. Bosch, X.; Hernández, C.; Pericas, J.M.; Doti, P.; Marušić, A.; Manzoli, L. Misconduct policies in high-impact biomedical journals. *PLoS One* 2012, *7*, e51928.

24. Resnik, D.; Peddada, S.; Brunson, W. Research misconduct policies of scientific journals. *Account. Res.* 2009, *16*, 254–267.

25. Resnik, D.; Patrone, D.; Peddada, S. Research Misconduct Policies of Social Science Journals and Impact Factor.*Account. Res.* 2010, *17*, 79–84.

26. Enders, W.; Hoover, G.A. Whose line is it? Plagiarism in economics. *J. Econ. Lit.* 2004, *42*, 487–493.

27. Wager, E.; Fiack, S.; Graf, C.; Robinson, A.; Rowlands, I. Science journal editors' views on publication ethics: Results of an international survey. *J. Med. Ethics* 2009, *35*, 348–353.

28. Williams, P.; Wager, E. Exploring Why and How Journal Editors Retract Articles: Findings from a Qualitative Study.*Sci. Eng. Ethics* 2013, *19*, 1–11.

29. Elia, N.; Wager, E.; Tramèr, M.R.; Wray, K.B. Fate of Articles That Warranted Retraction Due to Ethical Concerns: A Descriptive Cross-Sectional Study. *PLoS One* 2014, *9*, e85846.

30. Neale, A.V.; Dailey, R.K.; Abrams, J. Analysis of Citations to Biomedical Articles Affected by Scientific Misconduct. *Sci. Eng. Ethics* 2010, *16*, 251–261.

31. Neale, A.V.; Northrup, J.; Dailey, R.; Marks, E.; Abrams, J. Correction and Use of Biomedical Literature Affected by Scientific Misconduct. *Sci. Eng. Ethics* 2007, *13*, 5–24.

32. Resnik, D.B.; Dinse, G.E. Scientific retractions and corrections related to misconduct findings. *J. Med. Ethics* 2012, *39*, 46–50.

33. Lappi-Seppälä, T. Penal policy in Scandinavia. In *Crime, Punishment, and Politics in Comparative Perspective*; Tonry, M., Ed.; The University of Chicago Press: Chicago, IL, USA, 2007; Volume 36, pp. 217–295.

34. Lappi-Seppälä, T. Explaining imprisonment in Europe. *Eur. J. Criminol.* 2011, *8*, 303–328.

35. Heckman, J.J. Sample Selection Bias as a Specification Error. *Econometrica* 1979, *47*, 153–161.

36. Macilwain, C. Scientific Misconduct: More Cops, More Robbers? *Cell* 2012, *149*, 1417–1419.

37. Mongeon, P.; Larivière, V. The Collective Consequences of Scientific Fraud: An Anlysis of Biomedical Research. In Proceedings of the International Conference on Scientometrics and Informetretrics; Gorraiz, J., Ed.; Austrian Institut of Technology: Vienna, Austria, 2013; pp. 1897–1899.

38. Steen, R.G. Retractions in the scientific literature: Do authors deliberately commit research fraud? *J. Med. Ethics* 2011,*37*, 113–117.

39. Bilbrey, E.; O'Dell, N.; Creamer, J. A novel rubric for rating the quality of retraction notices. *Publications* 2014, *2*, 14–26.

40. Krohn, M.D.; Lizotte, A.J.; Phillips, M.D.; Thornberry, T.P.; Bell, K.A. Explaining Systematic Bias in Self-Reported Measures: Factors that Affect the Under- and Over-Reporting of Self-Reported Arrests. *Justice Q.* 2013, *30*, 501–528.

41. Maxfield, M.G.; Luntz Weiler, B.; Spatz Widom, C. Comparing Self-Reports and Official Records of Arrest. *J. Quant. Criminol.* 2000, *16*, 87–110.

42. Martinson, B.C.; Anderson, M.S.; de Vries, R. Scientists behaving badly. *Nature* 2005, *435*, 737–738.

43. Böhmer, S.; Neufeld, J.; Hinze, S.; Klode, C.; Hornbostel, S. *Wissenschaftlerbefragung 2010.Forschungsbedingungen von Professorinnen und Professoren in Deutschland*; iFQ-Working Paper No.8; Institut für Forschungsinformation und Qualitätssicherung: Bonn, Germany, 2011.

44. List, J.; Bailey, C.; Euzent, P.; Martin, T. Academic economists behaving badly? A survey on three areas of unethical behavior. *Econ. Inq.* 2001, *39*, 162–170.

45. Gardner, W.; Lidz, C.W.; Hartwig, K.C. Authors' reports about research integrity problems in clinical trials. *Contemp. Clin. Trials* 2005, *26*, 244–251.

46. Geggie, D. A survey of newly appointed consultants' attitudes towards research fraud. *J. Med. Ethics* 2001, *27*, 344–346.

47. Schwarz, N. Cognitive aspects of survey methodology. *Appl. Cogn. Psychol.* 2007, *21*, 277–287.

48. Becker, H.S. *Outsiders: Studies in the Sociology of Deviance*; Free Press: New York, NY, USA, 1973.

49. Frow, E.K. Drawing a line: Setting guidelines for digital image processing in scientific journal articles. *Soc. Stud. Sci.*2012, *42*, 369–392.

50. Vaughan, D. The Dark Side of Organizations: Mistake, Misconduct, and Disaster. *Annu. Revie. Sociol.* 1999, *25*, 271–305.

51. Knorr-Cetina, K. *Epistemic Cultures: How the Sciences Make Knowledge*; Harvard University Press: Cambridge, MA, USA, 1999.

52. Merton, R.K. *Social Theory and Social Structure*; Free Press: New York, NY, USA, 1968.

53. Collins, H.M. *Changing Order: Replication and Induction in Scientific Practice*; University of Chicago Press: Chicago, IL, USA, 1992.

54. Gieryn, T.F. Boundary-work and the demarcation of science from non-science: Strains and interests in professional ideologies of scientists. *Am. Sociol. Rev.* 1983, *48*, 781–795.

55. Gieryn, T.F.; Figert, A.E. Scientists protect their cognitive authority: The status degradation ceremony of Sir Cyril Burt. In *The Knowledge Society*; Böhme, G., Stehr, N., Eds.; Sociology of the Sciences; Springer: Dordrecht, The Netherlands, 1986; pp. 67–86.

56. Gieryn, T.F. *Cultural Boundaries of Science: Credibility on the Line*; University of Chicago Press: Chicago, IL, USA, 1999.

57. Flick, U. *Qualitative Sozialforschung: Eine Einführung; Rororo Rowohlts Enzyklopädie; Orig.-Ausg., vollst. überarb. und erw. Neuausg*; Rowohlt-Taschenbuch-Verl.: Reinbek bei Hamburg, Germany, 2007; Volume 55694.

58. Denzin, N.K.; Lincoln, Y.S. *The SAGE Handbook of Qualitative Research*, 3rd ed.; Sage Publications: Thousand Oaks, CA, USA, 2005.

59. King, G.; Keohane, R.O.; Verba, S. *Designing Social Inquiry: Scientific Inference in Qualitative Research*; Princeton Paperbacks; Princeton University Press: Princeton, NJ, USA, 1994.

60. Pinch, T.J. Normal Explanations of The Paranormal: The Demarcation Problem and Fraud in Parapsychology. *Soc. Stud. Sci.* **1979**, *9*, 329–348.

Chapter 11

SURVEY OF GREEN RADIO COMMUNICATIONS NETWORKS: TECHNIQUES AND RECENT ADVANCES

Mohammed H. Alsharif, Rosdiadee Nordin, and Mahamod Ismail

Department of Electrical, Electronics and Systems Engineering, Faculty of Engineering and Built Environment, Universiti Kebangsaan Malaysia, 43600 Bangi, Selangor, Malaysia

ABSTRACT

Energy efficiency in cellular networks has received significant attention from both academia and industry because of the importance of reducing the operational expenditures and maintaining the profitability of cellular networks, in addition to making these networks "greener." Because the base station is the primary energy consumer in the network, efforts have been made to study base station energy consumption and to find ways to improve energy efficiency. In this paper, we present a brief review of the techniques that have been used recently to improve energy efficiency, such as energy-efficient power amplifier techniques, time-domain techniques, cell switching, management of the physical layer through multiple-input multiple-output (MIMO) management, heterogeneous network architectures based on Micro-Pico-Femtocells, cell zooming, and relay techniques. In addition, this paper discusses the advantages and disadvantages of each technique to contribute to a better understanding of each of the techniques and thereby offer clear insights to researchers about how to choose the best ways to reduce energy consumption in future green radio networks.

INTRODUCTION

In the past few years, the cellular network sector has developed rapidly. This rapid growth is due to the increases in the numbers of mobile subscribers, multimedia applications, and data rates. According to [1], the data transmission rate doubles by a factor of approximately ten every five years. Figure 1 shows how the number of subscribers in cellular networks have increased [2].

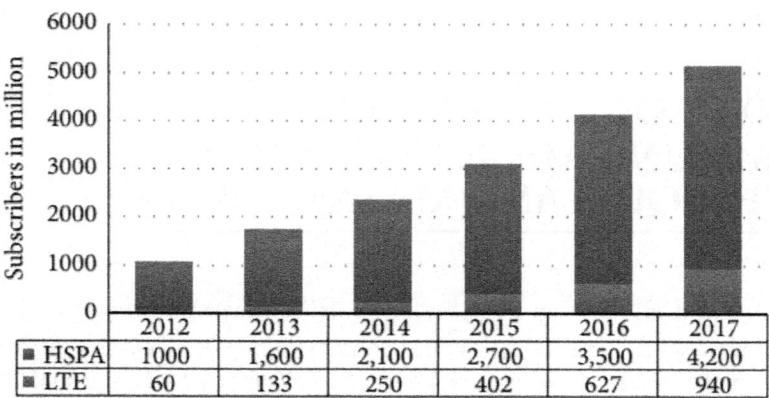

Figure 1: Growth forecasts for global HSPA and LTE subscribers, 2012–2017 [2].

The increase in the number of mobile subscribers has led to an increase in data traffic; as a result, the number of base stations (BSs) has increased to meet the needs of customers. Reference [3] describes the growth in the number of BSs in developing regions between 2007 and 2012, and forecasted that the total number of BSs would increase by over 2 million within this period. Most of the previous studies on this subject have focused on improving both system capacity and data rates, while neglecting the increasing demand of cellular networks for energy. This increasing energy demand has prompted considerable research on the subject of "green communications." This paper discusses the issue of energy efficiency in communications networks. Perhaps the two most important reasons to pursue the development of green communications networks are increases in carbon dioxide emissions (CO_2) and increases in operational expenditures (OPEX). CO_2 emissions are mainly associated with off-grid sites that provide coverage for remote areas. Most such sites are powered by diesel-power generators. According to [4], in 2002, the amount of CO_2 emissions associated with information and communication technology (ICT) was 151 Mt CO_2. The mobile communication sector was responsible for 43% of this total, and this proportion is expected to increase to 51% of the total, or 349 MtCO_2, by 2020. With respect to the economics of the sector, [5] indicates that ICT currently consumes 600 TWh (Terawatt hours) of electrical energy and that this consumption is expected to increase to 1,700 TWh by 2030. Cellular networks represent the largest component of the ICT sector. Figure 2 illustrates the total electricity consumption by communication networks around the world. Energy consumption by cellular networks is expected to increase rapidly in the future if no measures are taken to alter this trend [6].

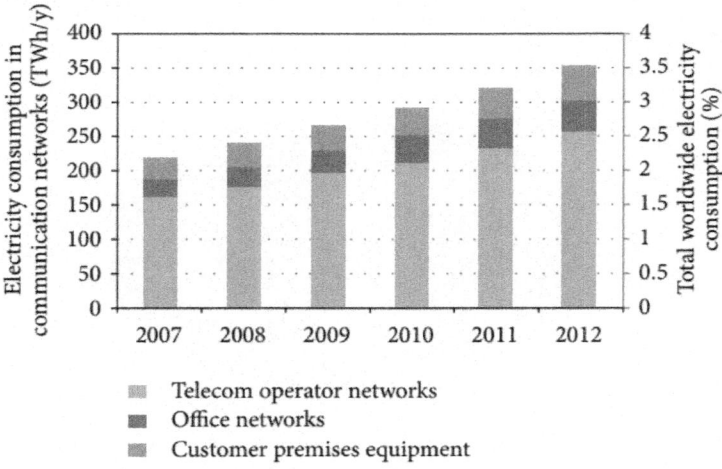

Telecom operator networks
Office networks
Customer premises equipment

Figure 2: Worldwide electricity consumption by communication networks [6].

The above-mentioned statistics have motivated researchers in both academia and industry to develop techniques to reduce the energy consumption of cellular networks, thereby maintaining profitability and making cellular networks "greener." Reference [7] highlighted the goals associated with green cellular networks:(i)improvement of energy efficiency,(ii)improvement of the intelligence of the network through tradeoffs between energy consumption and external conditions, that is, traffic loads,(iii)integration of the network infrastructure and network services to enable the network to be more responsive and to require less power to operate,(iv)reduced carbon emissions.

As shown in Figure 3, the BS is the main energy consumer in a cellular network [8]. Reference [3] indicated that the numbers of BSs are increasing and will continue to increase in the future to guarantee the quality of service (QoS) expected by mobile subscribers. As a result, energy consumption by base stations will continue to increase. Therefore, an effort is required to reduce the energy consumption of base stations, while continuing to provide the expected quality of service, taking into account the associated cost. The energy consumed by a BS consists of two components. The fixed component represents the energy consumed by the internal components of the BS, as shown in Figure 4, which require further classification of the components of BS sites and the energy consumption of each component. The dynamic component represents the energy consumed in RF transmission [9]. Thus, the solutions have two components: first, the hardware solution, for which the focus is on improving the energy consumption in the BS components, such as power amplifiers (PAs), digital signal processors (DSPs), cooling systems, and feeder cables and second, intelligent management of network elements based on traffic load variations.

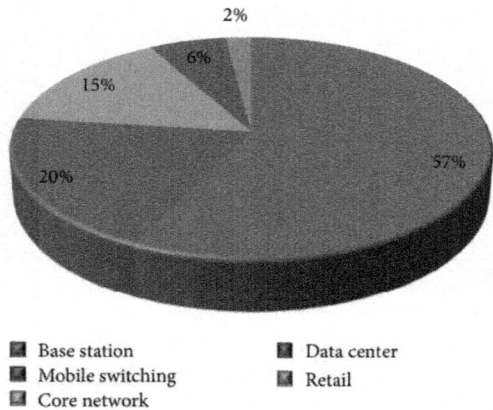

Figure 3: Energy consumption composition of a mobile operator.

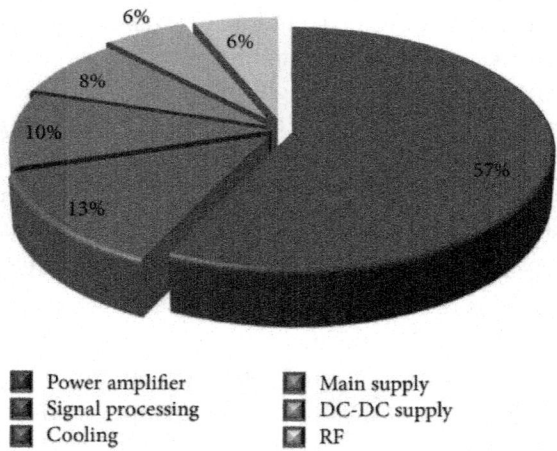

Figure 4: Distribution of energy consumption at a macrobase station site.

In this paper, we provide a brief overview of the techniques that have been considered in previous studies for use in saving energy, including a discussion of the principles of operation, the advantages, and the shortcomings of each technique. The graph shown in Figure 5 classifies the different techniques that will be discussed in the following sections.

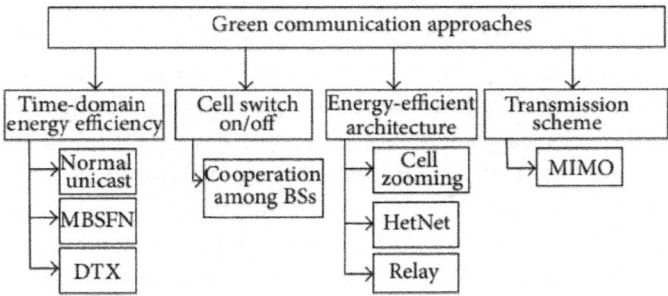

Figure 5: Classifications of energy-saving techniques.

The remainder of the paper is organised as follows. Section 2 discusses energy efficiency metrics. The energy consumption in the internal components of a BS, especially the power amplifier, is discussed in Section 3. Sections 4 through 7 address time-domain energy-efficient techniques, cell switch on/off techniques, energy-efficient network architectures, and transmission scheme. Finally, we compare the techniques considered in Section 8 and highlight each technique's contribution to energy-efficient radio communication.

ENERGY EFFICIENCY METRICS

Energy efficiency metrics provide information that can be used to assess and compare the energy consumption of various components of a cellular network and of the network as a whole. These metrics also help us to set long-term research goals for reducing energy consumption. With the increase in research activities pertaining to green communication and due to the intrinsic differences and relevance of various communication systems and performance measures, it is difficult for one single metric to suffice. Hence, several standardisation bodies and forums have considered energy efficiency in their network implementation strategies. However, energy efficiency metrics have been classified in three main categories in [10], that is, (i) facility-level, (ii) equipment-level, and (iii) network-level metrics. Reference [11], on the other hand, highlighted another type of metric, called the access-node level. The facility-level metric refers to high-level systems (such as data centres). The Green Grid (TGG) association of IT professionals proposed the metrics of power usage efficiency (PUE) and data centre efficiency (DCE) to evaluate energy efficiency in data centres [12]. Despite being a good metric for quickly assessing the performance of data centres at a macrolevel, PUE, which is defined as the ratio of total facility power consumption to total equipment power consumption, fails to account for the energy efficiency of individual pieces of equipment. Therefore, to quantify efficiency at the equipment level, a measure of the ratio of the energy consumption to the performance of a communication system would be more appropriate. Facility-level metrics assess initial power usage but do not reflect the energy efficiency of individual pieces of equipment. Thus, equipment-level metric, such as power amplifier efficiency metric, which quantify the performance of individual pieces of equipment, are required. The ATIS has introduced the telecommunications energy efficiency ratio (TEER), which is the ratio of useful work to power consumption and is measured in units of Gbps/Watt. Another equipment-level metric, the telecommunications equipment energy efficiency rating (TEEER), introduced by Verizon Networks and Building Systems, quantifies the total energy consumption as the weighted sum of the amounts of energy consumed by the equipment under different load conditions. Another equipment-level metric is the energy consumption rating (ECR), which is the ratio of the energy consumption to the effective system capacity, measured in units of Watt/Gbps [13]. However, even the busiest networks do not always operate under fullload conditions. Therefore, it would be useful to complement metrics such as the ECR to incorporate dynamic network conditions such as energy consumption under fullload, halfload, and idle conditions. Other metrics suitable for these purposes include the ECRW (weighted ECR), ECR-VL (energy efficiency over a variable-load cycle), and ECR-EX (energy efficiency over an extended-idle load cycle). Hence, ECR provides manufacturers insight into the performance of hardware components. However, these metrics (ECR, TEER, and TEEER) are unable to capture all the properties of a system. While the definitions of energy efficiency metrics at the component and equipment levels are fairly straightforward, it is more challenging to define energy efficiency metrics at the system or network level. Network-level metrics assess energy efficiency at the network level by considering the features and properties of the capacity and coverage of the network. The ETSI has defined two network-level energy efficiency metrics. The first metric is the ratio of the total coverage area to the power consumed at the site and is measured in units of km^2/Watt. The second metric is the ratio of the number of subscribers to the power consumed at the site and is measured in units of users/Watt [10]. Some specific metrics have been used to measure the performance of computing processing associated with energy consumption, in units such as millions of instructions per second per watt (MIPS/W) and millions of floating-point operations per second per watt (MFLOPS/W) [10]. Reference [14] highlighted a metric with units of energy per bit per unit area ($J/bit/m^2$). This metric relates energy consumption to the number of transferred bits and the area of coverage. This is equivalent to analysing the average power usage with respect to the average rate and the area of coverage (W/bps/m^2). A rich set of metrics exists at the access node level. The ECR quantifies the energy used to transmit a piece of information (Joules/bit). Some other metrics quantify the utility of various resources with respect to existing tradeoffs, such as the spectral efficiency (b/s/Hz) and the power efficiency (b/s/Hz/W). One metric intended to cover all the aspects in a more general way is the radio efficiency ((b·m)/s/Hz/W) [15], which reflects the data transmission rate and the transmission distance that is attainable for a given bandwidth and level of power supplied. To summarise the discussion above, a nonexhaustive list of energy metrics is given in Table 1. In addition, reference [16] discusses in detail the tradeoffs among several different energy efficiency metrics, such as deployment efficiency versus energy efficiency, spectrum efficiency versus energy efficiency, bandwidth versus power, and delay versus power. However, the most popular metric for measuring the performance of the system is "bits per Joule," which is the number of bits transmitted per Joule of energy. Interested readers can find a more comprehensive taxonomy of green metrics in [14, 17]. Reaching a consensus on a small set of standard energy metrics in future will not only accelerate research activities in green communications but also help to pave the way towards standardisation.

Table 1: Classification of some energy efficiency metrics.

Level	Units	Description
Facility level	Power usage efficiency is a ratio ≥ 1.	Defined as the ratio of total facility power consumption to total equipment power consumption
	Data centre efficiency is a percentage (%)	Defined as the reciprocal of PUE
Equipment level	PA efficiency is a ratio	The ratio of output to input power
	W/Gbps	The ratio of energy consumption to effective system capacity
	Gbps/W	The ratio of useful work to power consumption
	MIPS/W	Millions of instructions per second per Watt
	MFLOPS/W	Millions of floating-point operations per second per watt
Access node level	b/s/Hz	Spectral efficiency refers to the rate at which information can be transmitted over a given bandwidth in a specific communication system
	b/s/Hz/W	The spectral efficiency per Watt
	(b·m)/s/Hz/W	The radio efficiency measures the rate at which data are transmitted and the transmission distance attainable for a given bandwidth and power resources supplied
	J/bit	The energy consumption rating is defined as the number of bits transmitted per Joule of energy
Network level	km^2/W	The ratio of coverage area to site power consumption
	W/km^2	The power consumed per unit area
	Users/W	The ratio of users served during the peak traffic hour to site power consumption
	J/bit/m^2	The energy consumption with respect to the number of transferred bits and the coverage area
	W/bps/m^2	The average power usage with respect to the average transmission rate and the coverage area

POWER AMPLIFIER IMPROVEMENT

Several approaches to improving the energy efficiency of the internal components of a BS are discussed in [18,19]. These approaches include reducing the amount of energy consumed by cooling systems, feeder cables, and power amplifiers (PAs). PAs have attracted the largest share of attention in previous studies because they represent the greatest proportion of the energy consumption of BSs. In mobile communications, the power amplifier in a macrobase station consumes the most energy, 65% of the total energy consumed by all BS elements, as shown earlier in Figure 3. Consequently, highly efficient power amplifiers are essential to reducing OPEX costs for mobile network operators. The most efficient PA operating point is close to the maximum output power (near saturation). Unfortunately, nonlinear effects and OFDM modulation with nonconstant envelope signals force the power amplifier to operate in a more linear region, that is, 6 to 12 dB below saturation [20]. In this subsection, we review several methods that have been used to improve power amplifier energy efficiency. The first technique, known as Doherty designs [21], is a special design technique used to improve PA energy efficiency. The use of Doherty designs has improved energy efficiency by 30 to 35% over a narrow bandwidth [10]. The enhancement can be further improved to over 50% using a digital predistorted Doherty architecture and gallium nitride (GaN) amplifiers [22]. GaN is a special material used in the manufacture of PA transistors. The same energy efficiency enhancement has been achieved through the use of crest factor reduction (CFR) and digital predistortion (DPD) with Doherty PAs [18]. Reference [11] considered a class J amplifier, a switched-mode power amplifier (SMPA) and a drain modulation technique with higher energy efficiency than a predistorted Doherty amplifier and found that the class AB with digital predistortion improves the PA efficiency by 50%. Ghannouchi et al. [23] proposed an inverse class F PA designed for WiMAX applications at a carrier frequency within a range centred at approximately 2.45 GHz. A detailed analysis and discussion of the classes can be found in [24]. In addition, it has been noted that many factors must be taken into account in PA design, including the following:(i)high linearity, to satisfy higher-order modulation schemes,(ii)greater average output power levels,(iii)broader operating bandwidths, and(iv)OPEX reductions achieved by decreasing BS energy consumption.

At present, PAs operate on high levels of DC power supply, independent of the traffic load. Thus, for a major part of the day, power is wasted [20]. The reduction in power consumption can be addressed in two ways. The first approach defines the operating point adjustment of power amplifiers needed to minimise the power consumption at arbitrary signal levels, so that the power efficiency is optimised for low, medium, and high traffic loads [25]. The second approach involves the deactivation of power amplifier stages to save power during time slots without signal transmission. More detailed analysis and discussion can be found in [26].

Despite many efforts, hardware technologies for reducing energy consumption at the base station have not been able to achieve significant energy savings. Moreover, one cannot ignore the amount of energy that is wasted by inefficient utilisation of resources. These factors have led to a solution that utilises both equipment-level and network-level approaches. The network-level approaches seek to tune network-related parameters based on the sensing of external conditions, which enables the determination of the optimal transmission strategies for energy savings. The philosophy behind all the proposed methods is the same, reducing energy consumption based on the traffic load. In the following section, we will review how network-level approaches can help to improve the energy efficiency.

TIME DOMAIN ENERGY EFFICIENCY

Time-domain solutions seek to reduce the PA operating time by reducing control signals during low traffic or in the idle-case situation. Therefore, the amount of energy that can be saved using this approach depends on the PA offtime. In [27], the reduction in control signals with respect to the reference signals (RSs) was investigated. The uses of RSs in the network in this approach are listed as follows [28]:(i)UE channel estimation (coherent demodulation of downlink transmissions).(ii)UE cell-search measurements, both for neighbour-cell measurements for handover and for measurements before initial access.

In the long-term evolution (LTE) system, each frame includes 10 subframes (from 0 to 9). In [28], three ways to reduce the number of RSs are discussed: (i) normal unicast, (ii) multicast broadcast single-frequency networks (MBSFN), and (iii) cell discontinuous transmission (DTX), also known as the "microsleep" technique in some of the references in [29]. Normal unicast RSs, inserted at the 0th, 4th, 7th, and 11th OFDM symbols at every subframe, have been applied to subframes 1–4 and 6–9 and not to subframes 0 and 5, due to the need to transmit RSs and control signals, such as the primary synchronisation signal, the secondary synchronisation signal, and the broadcast channel control. The second form of RS, known as MBSF, seeks to further decrease the power amplifier transmission time by reducing the number of RSs to 1. If there are no active UEs in the idle traffic situation, there is no need to transmit RSs in subframes 1–4 and 6–9. This approach is the principle of DTX. Figure 6 illustrates the principles of these different techniques. In addition, Table 2 summarises the energy savings and on-time PA for each technique. Reference [30] determined some of the characteristics of these approaches, such as the impact on the specification, the impact on service, backward compatibility, and process time.

Table 2: Energy savings and on-time PA for different time domain techniques

Technique	On-time PA in a frame (%)	Energy savings (%)
Normal unicast	47	40
MBSFN	28	55
DTX	7.1	85

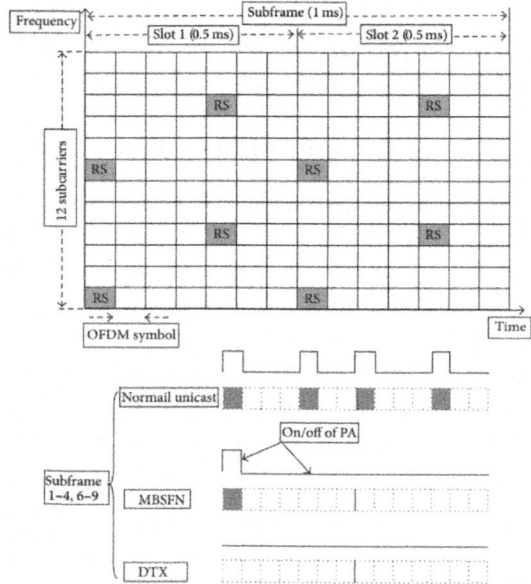

Figure 6: Time domain solution during idle and low traffic [27].

Similarly, in [31], switching off an optimum number of subframes per radio frame at low traffic loads was proposed.

Advantages. Clearly, this technique can result in a significant amount of energy savings in the idle traffic case. This means that it is an appropriate solution for rural areas. In addition, this technique saves energy in other components, such as ADC/DAC devices, DSP units, and cooling systems, during the switched-off periods.

Shortcomings. An insufficient quantity of RSs may cause some of the users to encounter problems during the synchronisation process with the BS, and thus, the UEs may be unable to enter into the DTX mode, thereby leading to a negative impact on the battery life of the UEs. In addition, some of the UEs may face difficulties in decoding the control signal.

CELL SWITCH ON/OFF

Switching off unused wireless resources and devices has become the most popular approach to reducing power consumption by cellular networks because it can save large amounts of energy. Cell switching is based on the traffic load conditions: if the traffic is low in a given area, some cells will be switched off, and the radio coverage and service will be provided by the remaining active cells. Therefore, the active cells will increase their transmission power to cover the area of the inactive cells. This may lead to a lack of coverage, because the BS maximum power is limited. A recent solution to this problem has been published in [32]. The authors propose an optimisation approach that achieves significant reductions in network energy consumption while abiding by the most important QoS constraints. The model proposed is a set of real-size universal mobile telecommunications system (UMTS) network instances consisting of various radio propagation environments. The results show that the proposed optimisation approach achieves significant reductions in network energy consumption (monthly energy savings between 35 and 57%) while respecting the most important QoS constraints, such as full area coverage and guaranteed service rates.

Switching cells off can be accomplished in both single-layer (macrocell) networks and multilayer network (such as HetNet). Reference [33] presents an overview of the energy savings achieved for a multilayer network. Obviously, not all the cells in a network have the same capability to execute switch off/on operations. Accordingly, the network sites are classified as coverage sites or capacity sites. Capacity booster cells can execute switch off/on operations in accordance with the operator's deployment policy, while coverage is guaranteed by coverage cells.

Conditions of Cell Switch off/on. Both eNB and operation, administration, and maintenance (OAM) systems can initiate cell switch off/on operations in response to various trigger mechanisms. The eNB depends on the real-time traffic load of its cells, while OAM systems depend on the historical statistics for the traffic load of a wide range of cells.

Cell Deactivation. The eNB will notify its neighbouring eNBs by X2 signalling to inform them that a specific cell will be switched off. The neighbouring eNBs will not allow their resident UEs to hand over control to that cell. In OAM systems, cell deactivation is initiated by explicit OAM commands for cell deactivation that are transmitted via Itf-N and then set down to the eNB. When the eNB successfully or unsuccessfully switches off the cell, it will return a response signal to the OAM system.

Cell Activation. It is initiated by the neighbouring eNB's trigger mechanism, which monitors the traffic load in the same area. An explicit X2 signalling message concerning cell activation is expected to request that the eNB that is in the cell deactivation state switches on to the original state again, that is, becomes active. Therefore, the neighbouring eNBs will allow their resident UEs to hand over control to this recovered cell again. For OAM, cell activation is similar to the cell deactivation that is initiated by the OAM's trigger mechanism. When the eNB successfully or unsuccessfully switches on the cell, it will return a response signal to the OAM system.

Reference [34] provides an overview of the cell on/off switch schemes or macrosleep in a single-layer network and describes how, when, and where to apply them in the scenario for rural macrocells at low traffic loads. It is important to keep in mind that in real networks, it is impossible to switch off a random number of BSs because switching off is restricted by their configuration. The energy savings in this approach depends on the number of cells that will be turned off. If a large number of cells are turned off, then the energy savings will be substantial. In addition, a tradeoff between the number of cells that will be turned off and the QoS should be considered, as discussed in [35]; accounting for this tradeoff it can achieve energy savings of 50%. The same authors, in [36], improved on their idea by introducing two different scenarios based on the UMTS network environment. The uniform scenario is based on the deployment of several identical microcells and a hierarchical scenario. In this scenario, an amendment to the topology of the network is implemented so that it contains one umbrella-cell overlap with many of the microcells. Marsan et al. demonstrated in [37] how to optimise energy savings by assuming that any fraction of cells can be switched off according to a deterministic traffic variation pattern over time and demonstrated that this approach can reduce energy consumption by between 25 and 30%. Reference [9] describes a study of the optimal number of active BSs that would be deployed based on the tradeoff between fixed power and dynamic power. Under various traffic load conditions, the energy savings can range from 12% to 40%. Reference [38] proposed the use of BSs with sleep/wake-up technology. A BS runs in active, sleep, or off mode depending on the traffic load. Each BS is augmented with an extra functional component, known as the "sleep/wake-up module," which is mainly used when the BS enters the sleep or wake-up mode. The energy consumption of the BS can be reduced by 72.9% using this strategy. Bousia et al. in [39] proposed a switching-off decision based on the average distance between BSs and UEs, whereby the BS at the maximum average distance will be switched off. This technique can achieve a reduction of up to 29% in energy consumption. Most of the previous studies that have discussed the energy efficiency problem have assumed that the conversion from switched on to switch off is instantaneous. Note that the switching period is very important because of its direct impact on the UEs in the cell. Reference [40] investigated the average time required for the implementation of the BS switch off by taking into account the handover process from the switched-off BS to a new BS. In addition, [30], the processing time, which takes between 50 and 100 ms of handover (HO) time, was highlighted. Reference [41] proposed three scalable BS switch-off patterns to reduce the power consumption of cellular networks during off-peak hours with the QoS of those switched-off cells guaranteed by focusing on the worst-case transmission/reception locations instead of calculating spatial averages. In addition, BS cooperation is used to efficiently extend the network coverage to the service areas of the switched-off cells. A potential power savings of up to 50% was observed in the numerical results.

Advantages: As mentioned earlier, the PA consumes the most power in a BS. Therefore, the cell switch-off approach achieves a good balance between performance, by providing coverage from neighbouring cells, and energy savings, by switching off some of the cells.

Shortcomings: Based on the traffic load, some cells are inactive and others are active; therefore, the active cells will increase their transmission power to cover the areas not covered by the inactive cells. There are several drawbacks to this method. First, the BS maximum power is limited; accordingly, there will be some areas without coverage, which can contribute to a decline in the quality of service. Second, the increase in power of the active cells reduces the energy savings. Finally, this approach reduces battery life for users because they require higher receiver power to connect with the other cells, which can be located at long distances away from them.

ENERGY-EFFICIENT ARCHITECTURE

This approach is considered a special case of the cell switch-off approach. The principle of this approach depends on the cooperation of the BSs to provide service to the users and provide energy savings in the event of low traffic. In this section, we present cell zooming, heterogeneous networks (HetNets), and relay techniques.

Cell Zooming

References [42, 43] proposed a cell capability to allow for the adjustment of the size of the cell according to the traffic load. When congestion occurs at the cell due to an increase in the number of UEs, the congested cell could zoom in, while the neighbouring cells with smaller amounts of traffic could zoom out to provide the coverage for UEs that cannot be served by the congested cell. If the neighbouring cells can provide coverage without the congested cell zooming in, then the congested cell can enter into sleep mode to reduce energy consumption. The major component of this design is a cell-zooming server (CS). The CS is considered the brain in this approach because it decides when cell zooming occurs. In addition, the CS collects network information, such as information on the traffic load, channel conditions, and user requirements. The CS analyses these data to decide whether to zoom. If a cell needs to zoom in or zoom out, the cell action is coordinated with the actions of its neighbouring cells with assistance from the CS. If the cell needs to zoom out, the transmission power of the BS will be increased, in contrast to the zoom in mechanism [42]. However, if the cells that must zoom out to provide coverage to the neighbouring cells that have switched off cannot provide this coverage due to the limited maximum transmission power of the BS, then an increase in energy consumption occurs because the neighbouring cells are unable to switch off, and there are areas without coverage, that is, coverage holes. In [44, 45], this problem was addressed through the deployment of more small cells to further improve energy consumption; that is, more cells can be switched off than in the traditional scheme. However, as the number of small cells increases, the fixed power of the BSs (e.g., the power required for cooling and for the power supply) also increases.

Advantages. This technique can improve the throughput and lengthen the UE's battery life.

Shortcomings. There are several drawbacks, as this physical adjustment technique depends on the transmission power of the BS. The BS's maximum transmission power is limited. In addition, when the remaining active cells increase their power during the zooming-out period, a problem of intercell interference emerges. Inter-cell interference occurs when all the neighbouring cells zoom out during the same time interval.

Heterogeneous Networks (HetNets)

Heterogeneous networks have a layered structure that combines different networks to serve the same mobile devices. Heterogeneous networks are intended to improve both throughput and energy consumption through the deployment of a network of cells of small size (such as micro-, pico-, and femtocells). These cells vary in size, output power, and data rate, as indicated in Table 3. Figure 7 illustrates the power consumption of the hardware elements in LTE BSs for different BS types.

Table 3: Cell size and transmitted power for a heterogeneous network [20]

Cell type	Cell size	Output power
Macro	1–30 km	Tens of watts
Micro	0.4–2 km	1–6.3 W
Pico	4–200 m	200 mW–2 W
Femto	10 m	20–200 mW

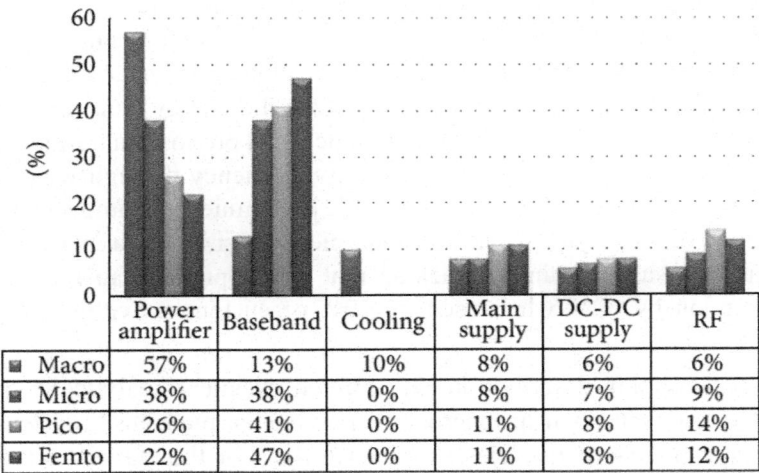

	Power amplifier	Baseband	Cooling	Main supply	DC-DC supply	RF
Macro	57%	13%	10%	8%	6%	6%
Micro	38%	38%	0%	8%	7%	9%
Pico	26%	41%	0%	11%	8%	14%
Femto	22%	47%	0%	11%	8%	12%

Figure 7: Power consumption of the hardware elements of LTE BSs for different BS types [20].

The mechanism of these networks is further discussed in [33]. Macrocells are deployed to provide overall coverage, while small cells become active if the demand increases. Based on this approach, a joint deployment strategy was investigated in [46] utilising microcells within a macrocell network, and the impact on energy consumption was determined. In addition, the authors considered area power consumption as a performance metric. The same authors, in [47], investigated the same issue in more detail. They evaluated and optimised the average number of microsites per macrocell. References [48, 49] investigated the energy efficiency of both homogeneous (pure microcells and macrocells) and heterogeneous networks consisting of a varying number of microsites, based on the traffic load conditions. The results of the study indicated that homogeneous microdeployment achieves better energy efficiency heterogeneous network deployment. The literature indicates that as the throughput and number of users in the network increase, the best approach to improving the energy efficiency of the network is to deploy more microsites. In addition to the above-mentioned studies, several papers have highlighted the feasibility of using pico- and femtocells to achieve energy savings. Femtocells are the subject of the largest share of these discussions in the literature. Femtocells are the closest to the users in terms of network size; that is, the distances are shorter, which results in less power transmission, thereby increasing energy efficiency and reducing the path loss. This reduced path loss improves the throughput of the network, as discussed in detail, along with other points, in [50]. Calin et al. [51] provided an overview of a joint deployment of femto- and macrocells. In addition, in [52, 53], the energy efficiency gains that can be achieved through the deployment of femtocells within a microcell, while accounting for the QoS, was discussed. The results indicated that the use of femtocells improved the energy consumption of the joint network deployment studied. Reference [54] proposed a new architecture called "FemtoWoC" to improve the energy efficiency through wireless-over-cable (WoC) transmission. Reference [22] examined the impact on energy efficiency of deployment of different picocell sizes in a macrocell. The simulation results obtained indicated that this approach can reduce energy consumption by 60% when used by 20% of customers within the picocell coverage area.

A recent study described in [55] offered excellent analytical models for the power consumption in macrocells, microcells, picocells, and femtocells. This paper discussed five classes of networks. For the class A network, the researchers considered a femtocell-based network in which, instead of macrocells, an area is fully covered by femtocells. The results showed that energy consumption was reduced by 82.72–88.37%. In the class B network, the coverage area was divided into three parts: an urban area, a suburban area, and a rural area, which were covered by femtocells, macrocells and portable femtocells, respectively. The simulation results obtained show that the total transmitted power was reduced by 78.53–80.19%. In the class C network, femtocells, picocells, microcells, and portable femtocells were allocated in a densely populated urban area, sparsely populated urban areas, suburban areas, and rural areas, respectively, which reduced the total transmitted power by 9.19–9.79%. In the class D network, microcells, picocells, and femtocells were allocated to border regions and macrocells were allocated to the rest of a coverage area. The simulation results obtained showed that the total transmitted power was reduced by 5.52–5.98%. In the class E network, femtocells were allocated to the boundary region of the macrocell and turned on in that region when the signal received from the macrocell BS was too low to successfully receive or generate a call. When all the femtocells were kept on, the macrocell shrinks in coverage area. The simulation results obtained showed that power consumption was reduced by 1.94–2.66%.

Advantages. A smaller cell size leads to improvement in coverage and capacity for the following reasons: lower transmission power, higher SINR, higher spectral efficiency, low path loss, prolonged handset lifetime (due to short transmit-receive distances), and smaller cells with lower costs.

Shortcomings. Despite the benefits that can be achieved by this technique, there are some challenges that need to be addressed, including management of the interfaces between heterogeneous environments and the dead zone problem.

Relay

Relay techniques are other means of saving energy, while improving network performance. The principle of this class of techniques is based on the deployment of relay nodes between the source (BS) and the destination (UEs). Relay nodes are intended to save energy, while increasing the network throughput by providing short transmission distances and thereby reducing path loss. In addition, the relay architecture reduces inter-cell interference due to low transmission power. There are two types of relay structures, namely, (i) pure relay systems and (ii) cooperative relay systems. A pure relay system is composed of small linked relay nodes between the BS and the UEs. Pure relay systems have been found to be more efficient than cooperative systems, and in some cases, a net power gain can be achieved for highly efficient relays. Additional benefits can be achieved when deploying relays to cover network traffic hotspots. In addition, managing sleep mode during MBSFN subframes is seen as a promising tool for mitigating the power consumption of relay nodes and increasing the benefits of their usage. Finally, we found that in urban and suburban environments, relay nodes can satisfy the needs of increasing traffic demand by providing higher energy efficiency without compromising the QoS and the system capacity [56]. In a cooperative relay system, several UEs act as the relay nodes and form a cooperative network. The cooperative relay technique provides path independence among different fading channels (a fundamental aspect of the diversity gain concept), thereby achieving energy savings [57]. Reference [56] highlighted the implementation of relay schemes for in-building deployment, which can increase energy efficiency in comparison to P2P communication. The literature suggests that the most important parameters in the design of relay networks are the following: (i) the choice of the relay node, (ii) the relay strategy, and (iii) the allocation of power and bandwidth for each user. Reference [58] discussed the optimal allocation of energy and bandwidth for each user and the selection of the optimal relay node and the optimal relay strategy (i.e., decode and forward versus amplify and forward), taking into account both user traffic demand and the physical channel. Reference [59] discussed several relay strategies, including the most significant ones: (i) amplify and forward (AF), (ii) decode and forward (DF), and (iii) compress and forward (CF). Reference [20] investigated hybrid relaying schemes, which allow the relay to dynamically switch between DF and CF schemes according to its decoding status. The analysis results showed that hybrid relaying achieves better energy efficiency performance than DF and offers the best energy efficiency performance, especially at the cell edge. Reference [60] discussed the impact on energy efficiency of the number of hops, the locations of the hops, the allocated power, and the data rate of each hop. Reference [61] investigated the improvement in energy efficiency of the users achieved by allocation of the optimal power to maximise the energy efficiency (EE) of each user. The authors of reference [62] proposed an energy-efficient single cooperative relay selection that accounts for both the MAC layer protocol and the power control strategy when selecting the node. Reference [63] investigated the energy consumption of single hops with fixed transmission power and multihops with fixed transmission power and power control in CDMA cellular networks. The results indicated that multihop with power control is better in terms of energy consumption.

Advantages. Whenever the distance between the transmitter and the receiver is small, this technique provides high coverage and capacity. The advantages of this technique are the following: low transmission power, low interference, reduced path loss, and extended handset life.

Shortcomings. There is a tradeoff in achieving energy efficiency in terms of the number of nodes that are required in a relay scheme and the optimal relay configuration. With respect to cooperative relay systems, the first challenge is that the radio frequency resources for each user should be split for transmitting data both from the user and from the other users, and the second issue is to select an appropriate user to act as a relay node.

TRANSMISSION SCHEME

Multiinput and multioutput (MIMO) has become the main feature of the evolution of the next generation of wireless networks. MIMO has the advantages of reducing fading and increasing throughput without it being necessary to increase either the bandwidth or the transmission power. These advantages can be achieved by introducing space-time coding (STC), which exploits spatial diversity to overcome fading by sending the signal that carries the same information through different paths, and by using spatial multiplexing (SM), which exploits multiple paths to send more information. In this subsection, we highlight the impact of MIMO transmission on energy efficiency. Reference [64] discussed a tradeoff between the diversity gain and the multiplexing gain in terms of the energy efficiency of wireless sensor networks. The discussion indicated that the energy efficiency of MIMO transmissions can be higher than that of single-input/single-output (SISO) transmission if the design is implemented properly. A significant energy reduction can be achieved if both diversity gain and multiplexing gain are used. The impact on energy efficiency of cooperative MIMO techniques with data aggregation for wireless sensor networks was discussed in [65], which demonstrated that energy consumption can be reduced by reducing the amount of data transmitted and by using an efficient resource allocation scheme. Cui et al. [66] presented an MIMO energy consumption model for both the transmitter and the receiver, based on Alamouti's diversity scheme. Using this model, the energy efficiency over various transmission distances was studied, and the energy efficiency values were compared with those obtained using a SISO system under the same conditions of throughput and BER. The results of the comparison suggests that MIMO is not always more energy efficient than SISO; indeed, at short distances, SISO may outperform MIMO in terms of energy efficiency. In addition, MIMO is not efficient for low traffic loads because it consumes more energy per circuit, as discussed in [67], in which switching between MIMO and SIMO under dynamic loads was studied. The results of the study indicated that SIMO achieves energy savings under low traffic loads. Some research studies have also focused on the traffic load adaptation of MIMO antennae [68, 69]. Huawei and Samsung discussed several methods for allowing the BS to change the number of active antennae based on the traffic load conditions. Another approach is to reduce the number of antennae based on the ratio of switch-off RF power amplifier units, for example, switching from 4-antennae to 1-antenna transmission to reduce energy consumption to 1/4 of the energy consumed in the 4-antennae case [70]. Each branch of a MIMO antenna is connected to a PA, and the PA is the component that consumes the most energy in a BS. Therefore, by using the adaptive MIMO approach, significant energy savings can be achieved during network operations. Details of the MIMO energy savings approach can be found in [69]. Reference [30] describes some of the benefits of this approach, such as improvements in energy savings, processing time, and the impact on network service.

Advantages: The MIMO scheme optimises the tradeoff between energy reduction and throughput, while reducing cochannel interference.

Shortcomings: There are some drawbacks to the reduced MIMO antenna approach, such as the need for a fixed antenna port during the active time of the BS and the requirement that the number of switched on/off antennae may be communicated to the UEs properly. Otherwise, the behaviour of the UEs will be impacted. In addition, a reduction in the number of antennae can result in service degradation or interruption during the antenna reconfiguration.

CONCLUSIONS AND COMPARISONS

This paper presents an overview of the energy consumption problems of wireless communication networks and describes the techniques that have been used to improve the energy efficiency of these networks. Table 4presents a summary of the techniques discussed in this paper. In addition, the table summarises the energy savings that can be achieved by each technique and other technical details that must be considered in improving energy efficiency.

Table 4: Evaluation of the energy-efficient solutions reviewed in this paper

Approach	Techniques		Energy savings/enhancements	Considerations	Comments
Power amplifier improvement	Digital predistorted Doherty-architectures and GaN [22]		Up to 50%	Linearity PAPR Cost	Improvement depends on a special design and material.
	CFR and DPD with Doherty [18, 22]				
	class-AB with digital predistortion [11]		Approximately 50%		
	class J amplifier [11]		70% to 90%		
	inverse class F [23]		74%		
	Switched-mode power amplifier (SMPA) [22, 24]		80–90%		
Time domain energy-efficient	Unicast [27]		40% to 50%	Synchronising UE battery life QoS	Energy savings depend on the PA operating time
	MBSFN [27, 30]		55%		
	DTX [28]		Up to 85%		
	Optimise subframes per frame [31]		90%		
Cell switch on/off	Increase ON cell radius [35, 36]	Residential scenario [35]	325%	Coverage UE battery life QoS	Energy savings depend on the number of BSs and the period of time over which each of the BSs is switched off
		Office scenario [35]	25–50%		
		Hierarchical scenario [35]	17%		
		Uniform scenario [36]	26–40.7%		
		Hierarchical scenario [36]	30%		
	Switch off any fraction of the cell according to a deterministic traffic scheme [37]		25–30%		
	Optimise the number of active BSs [9]		12–40%		
	Ecological protocooperation [38]		72.9%		
	Maximum average distance between BSs [39]		29%		
Energy-efficient architectures	Cell Zooming [42–45]		Up to 40%	Interference Dead zone problem	Must take into account interference management between heterogeneous environments
	HetNets [22, 46, 55]	Macrocell-microcell	44%		
		Macrocell-picocell	60%		
		Macrocell-femtocell	78–80%		
	Relay [57, 59]		8%–18%		
Transmission scheme	Reduce the number of antennae [69, 70]		50%, depending on the number of antennae.	QoS Impact on the UE	Energy savings depend on the number of antennae that will be switched off

Integration of multiple approaches based on changes in traffic load is possible and can help to maximise energy savings under low downlink traffic conditions. In addition, integrated solutions can combine component-, link- and network-level energy saving techniques. For instance, the cell switch-off technique with cooperation among mobile network providers is seen as having potential for implementation in industrial areas because the traffic loads in these areas are low for long periods of time, especially at night and over weekends and holidays. Preliminary studies have shown that this approach significantly improves energy savings, compared to a stand-alone solution. Nonetheless, there are issues that need additional investigation in future studies, such as resource management, increasing the load on the host network (due to its own customers plus roaming customers), and the ability to provide good service as well as handling of traffic data. Examples of such challenges include VoIP calls or video streaming with no delay, compatibility between signalling loads and channel capacity, and communication overhead due to handovers. Other important issues include the variation of mobile operator coverage and the types of services from one area to another. In addition, the tradeoff between computational complexity and the savings in required transmission power is an important issue. The complexity increases with the number of networks. Emphasis should be placed on designing low-complexity schemes, resulting in easy, viable implementations as well as gains in overall network-level energy efficiency.

In the same context, a BS can be configured to use a single antenna and the maximum number of MBSFN subframes in a frame. The challenges of this hybrid solution are the processing/interruption time and signalling for system reconfiguration and avoidance of adverse impacts on UE performance. The previously discussed hybrid approach of combining cell on/off switching with cell zooming can also be applied. Compatibility between the signalling load and channel capacity, intercell interference, coverage, and the maximum capacity achievable are all issues that need to be further studied and analysed thoroughly.

ACKNOWLEDGMENTS

The authors would like to thank the Universiti Kebangsaan Malaysia for its financial support of this work under Grant Ref: UKM-GUP-2011-065. The authors also would like to thank the anonymous reviewers for their feedback.

REFERENCES

1. G. Fettweis and E. Zimmermann, "ICT energy consumption-trends and challenges," in Proceedings of the 11th International Symposium on Wireless Personal Multimedia Communications (WPMC '08), pp. 1–6, 2008.

2. V. Livingston, "HSPA and LTE mobile broadband in the Americas," in Proceedings of the Workshop of 4G Americas, 2012.

3. Green Power for Mobile, GSMA, "Community Power Using Mobile to Extend the Grid," 2010,http://www.altobridge.com/wp-content/uploads/2010/01/Community-Power.pdf.

4. The Climate Group. SMART 2020, "Enabling the low carbon economy in the information age," 2008,http://www.smart2020.org/_assets/files/02_Smart2020Report.pdf.

5. I. Humar, X. Ge, L. Xiang, M. Jo, M. Chen, and J. Zhang, "Rethinking energy efficiency models of cellular networks with embodied energy," IEEE Network, vol. 25, no. 2, pp. 40–49, 2011.

6. S. Lambert, W. V. Heddeghem, W. Vereecken, B. Lannoo, D. Colle, and M. Pickavet, "Worldwide electricity consumption of communication networks," Optics Express, vol. 20, no. 26, pp. B513–B524, 2012.

7.   S. S. Sandhu, A. Rawal, P. Kaur, and N. Gupta, "Major components associated with green networking in information communication technology systems," in Proceedings of the International Conference on Computing, Communication and Applications (ICCCA '12), pp. 1–6, February 2012.

8.   C. Han, T. Harrold, S. Armour et al., "Green radio: radio techniques to enable energy-efficient wireless networks," IEEE Communications Magazine, vol. 49, no. 6, pp. 46–54, 2011.

9.   L. Xiang, F. Pantisano, R. Verdone, X. Ge, and M. Chen, "Adaptive traffic load-balancing for green cellular networks," in Proceedings of the IEEE 22nd International Symposium on Personal, Indoor and Mobile Radio Communications (PIMRC '11), pp. 41–45, September 2011.

10. T. Chen, H. Kim, and Y. Yang, "Energy efficiency metrics for green wireless communications," inProceedings of the International Conference on Wireless Communications and Signal Processing (WCSP '10), pp. 1–6, October 2010.

11. L. Suarez, L. Nuaymi, and J.-M. Bonnin, "An overview and classification of research approaches in green wireless networks," EURASIP Journal on Wireless Communications and Networking (WCN), vol. 2012, article 142, 2012.

12. C. Belady, A. Rawson, J. Pfleuger, and T. Cader, "Green grid data center power efficiency metric: PUE and DCIE," The Green Grid, 2008.

13. A. Amanna, Green Communications, Institute for Critical Technology and Applied Science (ICTAS) at Virginia tech, 2009.

14. M. A. Imran, J. Rubio, G. Auer, et al., "Most suitable efficiency metrics and utility functions," EARTH Project Report, Deliverable D2.4, pp. 1–89, 2012.

15. L. Zhao, J. Cai, and H. Zhang, "Radio-efficient adaptive modulation and coding: green communication perspective," in Proceedings of the IEEE 73rd Vehicular Technology Conference (VTC '11), pp. 1–5, May 2011.

16. Y. Chen, S. Zhang, S. Xu, and G. Y. Li, "Fundamental trade-offs on green wireless networks," IEEE Communications Magazine, vol. 49, no. 6, pp. 30–37, 2011.

17. H. Hamdoun, P. Loskot, T. O'Farrell, and J. He, "Survey and applications of standardized energy metrics to mobile networks," Annales des Telecommunications/Annals of Telecommunications, vol. 67, no. 3-4, pp. 113–123, 2012.

18. ATIS Exploratory Group on Green (EGG), "ATIS report on wireless network energy efficiency,"Alliance for Telecommunications Industry Solutions, 2010.

19. Mobile VCE, "Power Amplifiers for 4G and beyond—managing the efficiency, bandwidth and linearity tradeoff," http://www.mobilevce. com/sites/default/files/infostore/GR%20POWER%20AMP.pdf.

20. G. Auer, O. Blume, V. Giannini, et al., "Energy efficiency analysis of the reference systems, areas of improvements and target breakdown," EARTH Project Report, Deliverable D2.3, pp. 1–68, 2012.

21. H. F. Raab, "Efficiency of doherty RF power-amplifier systems," IEEE Transactions on Broadcasting, vol. 33, no. 3, pp. 77–83, 1987.

22. H. Claussen, L. T. W. Ho, and F. Pivit, "Effects of joint macrocell and residential picocell deployment on the network energy efficiency," in Proceedings of the IEEE 19th International Symposium on Personal, Indoor and Mobile Radio Communications (PIMRC '08), pp. 1–6, September 2008.

23. F. M. Ghannouchi, M. M. Ebrahimi, and M. Helaoui, "Inverse class f power amplifier for WiMAX applications with 74% efficiency at 2.45 GHz," in Proceedings of the IEEE International Conference on Communications Workshops (ICC '09), pp. 1–5, June 2009.

24. B. Berglund, J. Johansson, and T. Lejon, "High efficiency power amplifiers," Ericsson Review, vol. 83, no. 3, pp. 92–96, 2006.

25. D. Ferling, A. Ambrosy, S. Petersson, et al., "Final report on green radio technologies," EARTH Project Report, Deliverable D4.3, pp. 1–121, 2012.

26. D. Ferling, A. Ambrosy, S. Petersson, et al., "Green radio technologies," EARTH Project Report, Deliverable D4.2, pp. 1–89, 2012.

27. T. Chen, Y. Yang, H. Zhang, H. Kim, and K. Horneman, "Network energy saving technologies for green wireless access networks," IEEE Wireless Communications, vol. 18, no. 5, pp. 30–38, 2011.

28. 3GPP R1-100387, "Extended cell DTX," 2010,http://www.3gpp.org/ftp/ tsg_ran/WG1_RL1/TSGR1_59b/Docs/R1-100387.zip.

29. M. Imran, A. Ambrosy, O. Blume, et al., "Final integrated concept," EARTH Project Report, Deliverable D6.4, pp. 1–95, 2012.

30. 3GPP R2-101213, "Energy saving techniques for LTE," 2010,http:// www.3gpp.org/ftp/tsg_ran/WG2_RL2/TSGR2_69/docs/R2-101213.zip.

31. R. Wang, J. S. Thompson, and H. Haas, "A novel time-domain sleep mode design for energy-efficient LTE," in Proceedings of the 4th International

Symposium on Communications, Control, and Signal Processing (ISCCSP '10), pp. 1–4, March 2010.

32. J. Lorincz, A. Capone, and D. Begusic, "Impact of service rates and base station switching granularity on energy consumption of cellular networks," EURASIP Journal on Wireless Communications and Networking, vol. 2012, 342, 2012.

33. 3GPP R3-100162, "Overview to LTE energy saving solutions to cell switch off/on," 2010,http://www.3gpp.org/ftp/tsg_ran/WG3_Iu/TSGR3_66bis/docs/R3-100162.zip.

34. E. Calvanese-Strinati, M. Kamoun, and M. Sarkiss, "Green network technologies," EARTH Project Report, Deliverable D3.2, pp. 1–101, 2012.

35. L. Chiaraviglio, D. Ciullo, M. Meo, and M. A. Marsan, "Energy-aware UMTS access networks," inProceedings of the 11th International Conference on Wireless Personal Multimedia Communications (WPMC '08), pp. 1–8, 2008.

36. L. Chiaraviglio, D. Ciullo, M. Meo, and M. A. Marsan, "Energy-efficient management of UMTS access networks," in Proceedings of the 21st International Teletraffic Congress (ITC '09), pp. 1–8, September 2009.

37. M. A. Marsan, L. Chiaraviglio, D. Ciullo, and M. Meo, "Optimal energy savings in cellular access networks," in Proceedings of the IEEE International Conference on Communications Workshops (ICC '09), pp. 1–5, June 2009.

38. M. F. Hossain, K. S. Munasinghe, and A. Jamalipour, "A protocooperation-based sleep-wake architecture for next generation green cellular access networks," in Proceedings of the 4th International Conference on Signal Processing and Communication Systems (ICSPCS '2010), pp. 1–8, December 2010.

39. A. Bousia, A. Antonopoulos, L. Alonso, and C. Verikoukis, "'Green' distance-aware base station sleeping algorithm in LTE-advanced," in Proceedings of the IEEE International Conference on Communications (ICC '12), pp. 1–5, 2012.

40. M. A. Marsan, L. Chiaraviglio, D. Ciullo, and M. Meo, "Switch-off transients in cellular access networks with sleep modes," in Proceedings of the IEEE International Conference on Communications Workshops (ICC '11), pp. 1–6, June 2011.

41. F. Han, Z. Safar, S. W. Lin, Y. Chen, and K. J. R. Liu, "Energy-efficient cellular network operation via base station cooperation," in Proceedings

of the IEEE International Conference on Communications (ICC '12), pp. 4374–4378, 2012.

42. Z. Niu, Y. Wu, J. Gong, and Z. Yang, "Cell zooming for cost-efficient green cellular networks," IEEE Communications Magazine, vol. 48, no. 11, pp. 74–79, 2010. · ·

43. R. Balasubramaniam, Cell zooming techniques for power efficient base station operation [M.S. thesis], Electrical Engineering, San Diego State University, 2012.

44. X. Weng, D. Cao, and Z. Niu, "Energy-efficient cellular network planning under insufficient cell zooming," in Proceedings of the IEEE 73rd Vehicular Technology Conference (VTC '11), pp. 1–5, May 2011.

45. Z. Niu, "TANGO: traffic-aware network planning and green operation," IEEE Wireless Communications, vol. 18, no. 5, pp. 25–29, 2011.

46. F. Richter, A. J. Fehske, and G. P. Fettweis, "Energy efficiency aspects of base station deployment strategies for cellular networks," in Proceedings of the IEEE 70th Vehicular Technology Conference Fall (VTC '09), pp. 1–5, September 2009.

47. A. J. Fehske, F. Richter, and G. P. Fettweis, "Energy efficiency improvements through micro sites in cellular mobile radio networks," in Proceedings of the IEEE Globecom Workshops, pp. 1–5, December 2009.

48. F. Richter and G. Fettweis, "Cellular mobile network densification utilizing micro base stations," inProceedings of the IEEE International Conference on Communications (ICC '10), pp. 1–5, May 2010.

49. F. Richter, A. J. Fehske, M. Patrick, and G. P. Fettweis, "Traffic demand and energy efficiency in heterogeneous cellular mobile radio networks," in Proceedings of the 71st IEEE Vehicular Technology Conference (VTC '10), pp. 1–6, 2010.

50. V. Chandrasekhar, J. G. Andrews, and A. Gatherer, "Femtocell networks: a survey," IEEE Communications Magazine, vol. 46, no. 9, pp. 59–67, 2008.

51. D. Calin, H. Claussen, and H. Uzunalioglu, "On femto deployment architectures and macrocell offloading benefits in joint macro-femto deployments," IEEE Communications Magazine, vol. 48, no. 1, pp. 26–32, 2010.

52. Y. Hou and D. I. Laurenson, "Energy efficiency of high QoS heterogeneous wireless communication network," in Proceedings of the IEEE 72nd

Vehicular Technology Conference Fall (VTC '10), pp. 1–5, September 2010.

53. F. Cao and Z. Fan, "The tradeoff between energy efficiency and system performance of femtocell deployment," in Proceedings of the 7th International Symposium on Wireless Communication Systems (ISWCS '10), pp. 315–319, September 2010.

54. J. Gambini and U. Spagnolini, "Wireless over cable for energy-efficient femtocell systems," inProceedings of the IEEE Globecom Workshops (GC '10), pp. 1464–1468, December 2010. · ·

55. A. Mukherjee, S. Bhat t acherjee, S. Pal, and D. De, "Femt ocell based green power consumpt ion met hods for mobile net work," Computer Networks, vol. 57, no. 1, pp. 62–178, 2013.

56. I. Gódor, L. Hévizi, O. Blume, et al., "Final report on green network technologies," EARTH Project Report, Deliverable D3.3, pp. 1–131, 2012.

57. G. Y. Li, Z. Xu, C. Xiong et al., "Energy-efficient wireless communications: tutorial, survey, and open issues," IEEE Wireless Communications, vol. 18, no. 6, pp. 28–35, 2011.

58. T. C.-Y. Ng and W. Yu, "Joint optimization of relay strategies and resource allocations in cooperative cellular networks," IEEE Journal on Selected Areas in Communications, vol. 25, no. 2, pp. 328–339, 2007.

59. Y. Yang, H. Hu, J. Xu, and G. Mao, "Relay technologies for WiMAX and LTE-advanced mobile systems," IEEE Communications Magazine, vol. 47, no. 10, pp. 100–105, 2009.

60. C. Bae and W. E. Stark, "Energy-bandwidth tradeoff with spatial reuse in wireless multi-hop networks," in Proceedings of the IEEE Military Communications Conference (MILCOM '08), pp. 1–7, November 2008.

61. M. Nokleby and B. Aazhang, "User cooperation for energy-efficient cellular communications," inProceedings of the IEEE International Conference on Communications (ICC '10), pp. 1–5, May 2010.

62. Z. Zhou, S. Zhou, J.-H. Cui, and S. Cui, "Energy-efficient cooperative communication based on power control and selective single-relay in wireless sensor networks," IEEE Transactions on Wireless Communications, vol. 7, no. 8, pp. 3066–3079, 2008.

63. A. Radwan and H. S. Hassanein, "Does multi-hop communication extend the battery life of mobile terminals?" in Proceedings of the IEEE GLOBECOM, pp. 1–5, December 2006.

64. W. Liu, X. Li, and M. Chen, "Energy efficiency of MIMO transmissions

in wireless sensor networks with diversity and multiplexing gains," in Proceedings of the IEEE International Conference on Acoustics, Speech, and Signal Processing (ICASSP '05), pp. IV897–IV900, March 2005.

65. Y. Gai, L. Zhang, and X. Shan, "Energy efficiency of cooperative MIMO with data aggregation in wireless sensor networks," in Proceedings of the IEEE Wireless Communications and Networking Conference (WCNC '07), pp. 792–797, March 2007.

66. S. Cui, A. J. Goldsmith, and A. Bahai, "Energy-efficiency of MIMO and cooperative MIMO techniques in sensor networks," IEEE Journal on Selected Areas in Communications, vol. 22, no. 6, pp. 1089–1098, 2004.

67. H. Kim, C.-B. Chae, G. De Veciana, and R. W. Heath Jr., "A cross-layer approach to energy efficiency for adaptive MIMO systems exploiting spare capacity," IEEE Transactions on Wireless Communications, vol. 8, no. 8, pp. 4264–4275, 2009.

68. 3GPP R2-094677, "eNB power saving by changing antenna number," 2009,http://www.3gpp.org/ftp/tsg_ran/WG2_RL2/TSGR2_67/docs/R2-094677.zip.

69. 3GPP R2-094851, "Number of antennas change," 2009,http://www.3gpp.org/ftp/tsg_ran/WG2_RL2/TSGR2_67/docs/R2-094851.zip.

70. T. Chen, H. Zhang, Z. Zhao, and X. Chen, "Towards green wireless access networks," in Proceedings of the 5th International ICST Conference on Communications and Networking in China (ChinaCom '10), pp. 1–6, August 2010.

CITATION

CHAPTER 1

A. Ghassemi and T. A. Gulliver, "Low-Complexity Distortionless Techniques for Peak Power Reduction in OFDM Communication Systems," Journal of Computer Networks and Communications, vol. 2012, Article ID 929763, 13 pages, 2012. doi:10.1155/2012/929763

CHAPTER 2

C. Tselios, C. Papageorgiou, K. Birkos, I. Politis, and T. Dagiuklas, "Real-Time Communications in Autonomic Networks: System Implementation and Performance Evaluation," Journal of Computer Networks and Communications, vol. 2012, Article ID 485875, 10 pages, 2012. doi:10.1155/2012/485875.

CHAPTER 3

E. Barka and A. Lakas, "Integrating Usage Control with SIP-Based Communications," Journal of Computer Systems, Networks, and Communications, vol. 2008, Article ID 380468, 8 pages, 2008. doi:10.1155/2008/380468.

CHAPTER 4

Aohan Li, Ziheng Yang, Renji Qi, Feng Zhou and Guangjie Han, Code Synchronization Algorithm Based on Segment Correlation in Spread Spectrum Communication, doi:10.3390/a8040870.

CHAPTER 5

Andres Diaz Lantada, Carlos González Bris, Pilar Lafont Morgado and Jesús Sanz Maudes, Novel System for Bite-Force Sensing and Monitoring Based on Magnetic Near Field Communication, doi:10.3390/s120911544.

CHAPTER 6

Juan Chóliz, Ángela Hernández and Antonio Valdovinos, A Framework for UWB-Based Communication and Location Tracking Systems for Wireless Sensor Networks, doi:10.3390/s110909045.

CHAPTER 7

Gao Zhiqiang, Associate Professo, Design of CMOS Integrated Q-enhanced RF Filters for Multi-Band/Mode Wireless Applications, http://cdn.intechweb.org/pdfs/14268.pdf

CHAPTER 8

Juan-de-Dios Sánchez- López, Arturo Arvizu M, Francisco J. Mendieta and Iván Nieto Hipólito (2011). Trends of the Optical Wireless Communications, Advanced Trends in Wireless Communications, Dr. Mutamed Khatib (Ed.), ISBN: 978-953-307-183-1, InTech, DOI: 10.5772/15493.

CHAPTER 9

Robert Nagel and Stefan Morscher (2011). Connectivity Prediction in Mobile Vehicular Environments Backed By Digital Maps, Advanced Trends in Wireless Communications, Dr. Mutamed Khatib (Ed.), ISBN: 978-953-307-183-1, InTech, DOI: 10.5772/15991.

CHAPTER 10

Felicitas Hesselmann, Verena Wienefoet,and Martin Reinhart, Measuring Scientific Misconduct—Lessons from Criminology, doi:10.3390/publications2030061

CHAPTER 11

Mohammed H. Alsharif, Rosdiadee Nordin, and Mahamod Ismail, "Survey of Green Radio Communications Networks: Techniques and Recent Advances," Journal of Computer Networks and Communications, vol. 2013, Article ID 453893, 13 pages, 2013. doi:10.1155/2013/453893.

INDEX